The Ancient World in Motion

The Ancient World in Motion

Wheels, Roads, Ships, and the Technologies That Connected the Ancient World

Charles Kensington

Hidden Histories Media

Staunton, VA 24401

Printed in the United States of America

First Printing, 2025

Published by Hidden Histories Media

Paperback ISBN: 978-1-971137-08-7

Contents

Introduction

Ancient wheel ruts lie sunken beneath the asphalt of a modern highway in Turkey. Cars roar overhead at speeds no horse could match, but the ground remembers. If you walk the quiet shoulder after rain, the lines in the earth appear like ghosts, deep grooves carved by wooden wheels thousands of years ago. Follow them a short distance, and the trail disappears under smooth pavement that leads to a ferry terminal where ships glide across the sea guided by radar and satellite. Travelers buy tickets on glowing screens. Shipping containers stack like bright metal fortresses. Engines thunder into motion. Yet every modern movement here, every turn of a tire and shift of a rudder, follows a path first imagined by people who used rope, stone, and courage to cross the world.

Across the planet, modern mobility conceals ancient roots. Beneath airports sit the foundations of forgotten harbors. Beneath bullet trains lie the bones of Roman roads. Beneath fiber-optic cables runs the memory of postal riders who once galloped through the night to deliver a king's command before danger could spread. We are racing forward while walking the same old trails.

Why do humans work so hard to connect distant places and distant lives? Why risk oceans, deserts, mountains, and enemies for the chance to reach one another? Movement has always defined us.

A World Built on Motion

Imagine a family in ancient Mesopotamia tugging a sled of heavy clay pots toward a riverbank, watching animals strain and grit their teeth. Then picture the moment someone rounds a slab of wood into a circle, places it beneath the burden, and glimpses a future rolling toward them. That first wheel changed everything.

Soon, oxen pulled carts. Traders ventured farther. Cities grew. Distance shrank. Then came chariots thundering across plains, ships skimming coastlines, camel caravans disappearing into dunes, and double-hulled canoes gliding across the open Pacific guided by stars and memory. Empires rose on roads and fell when those roads crumbled. Libraries filled because couriers ran relay after relay, carrying scrolls to hungry minds before war or fire stole them away.

Civilization did not begin with writing or agriculture alone. It began when we discovered how to move, faster, farther, together.

Curiosity as Compass

People have always been drawn to what lies beyond the horizon. The need feels instinctive, as if our feet refuse to stay still. Curiosity pushes us out the door. Trade ensures we return with something worth carrying home. And communication keeps our world from splintering apart.

Motion spreads opportunity. Farmers reach new markets. Sailors find new spices. Astronomers discover new stars. Ideas that might have died in isolation become sparks that light entire eras.

Yet every advance also demanded risk. Pirates lurked behind headlands. Storms swallowed fleets without warning. Bridges collapsed under armies that depended on them. Whenever societies connected, both prosperity and vulnerability traveled the same road.

Even today, a disruption in shipping sends ripples through global economies. A blocked canal or broken supply chain reminds us that mobility remains our strength, and our weakness.

The Roads Beneath our Feet

What if the history of technology is not a list of inventions, but a story about connection? Wheels, roads, ships, bridges, compasses, maps, none exist for their own sake. They exist to bring people, goods, and knowledge together in ways that redefine what is possible.

In this book, you will step into clay workshops as the first wheel turns under human hands. You will stand on the deck of a trireme as rowers dig into the sea and cities rise in the distance. You will walk the Persian Royal Road with a courier switching horses at a relay station, racing to deliver a message that could change an empire. You will smell the smoke of blacksmith forges where metal bits and stirrups reshape armies. You will look down from rope bridges swinging above Andean rivers and feel the pride of Inca engineers who refused to let geography conquer them.

You will witness how every advance in motion, across land or sea, reshaped the world and every society within it.

This is not a survey of machines. It is the story of humanity refusing to remain apart.

When Movement Changes Everything

Across each chapter, a pattern emerges.
A new technology appears.
Distance loses power.
Culture transforms.

Roads do not merely guide feet. They influence language, identity, and power. When ideas circulate freely, societies flourish. When movement collapses, so does everything built upon it.

After Rome fell, fields encroached where highways once shined. Isolation returned. Innovation slowed. The world felt larger again, not because the land grew, but because access shrank.

Today, the world feels small because ancient people laid the foundation for connection. Their breakthroughs echo through GPS satellites, shipping lanes, airline routes, and data networks that circle the planet like invisible rivers. Even the internet depends on physical pathways of cables that mirror ancient trade corridors.

Nothing about mobility today is new, not the ambition behind it, and not the fragility that comes when its systems fail.

The Eternal Journey

Look again at the satellite image. The lights of Europe form a lattice that traces Roman roads almost perfectly. The Red Sea glows with shipping vessels following monsoon routes first mastered by ancient sailors. The Silk Road's spine remains visible in cities that still thrive where merchants once bartered under desert stars.

The technology has changed. The purpose has not. We move because movement expands what it means to be human. It brings comfort and challenge, ideas and conflict, progress and danger. It allows us to know one another, to rely on one another, to shape a future greater than what any city or kingdom could achieve alone.

The ancient world did not merely invent transportation. It invented connection. That invention never ended. It only accelerated.

Let the Journey Continue

What follows is a story of courage, brilliance, and persistence. It is a story written in stone roads and open sea, in hoofbeats and sailcloth, in messages passed from hand to hand under urgent skies. It is our story.

Turn the page and step into the workshop where a potter's wheel spins under glimmering sunlight.

Watch the first wheel rise, wobble, and roll, and feel the world beginning to move.

Chapter 1
The First Turning –
The Invention of the Wheel

The pottery workshop was warm and noisy with steady movement. In ancient Mesopotamia around 3500 BCE, a potter leaned over a spinning round platform, guiding the shape of a clay bowl with careful fingers. Sunlight slipped through the reed-mat doorway, forming a bright stripe across the turning disk. Outside, donkeys brayed and goats called across the settlement. Farmers shouted as they prepared for a long day in the fields. Inside the workshop, though, the world felt focused on one simple action: a circle turning on its center.

The potter's wheel was already a major innovation. It helped workers make more pottery faster than shaping each piece by hand. But no one yet imagined attaching that wheel to something that could move across the earth. Its circular motion was locked in place, fixed to the ground. Still, the idea of rotation was planted. It was only a matter of time before someone asked a powerful question: What if this could roll?

Soon, that question would lead to one of the most important inventions in human history. It would change how people moved food, tools, building supplies, and even themselves. The turning clay platform inside that humble workshop held the first small hint that the world could move in circles.

From Potters to Carts

Before the wheel existed, heavy loads were dragged on wooden sledges. Workers hauled goods with raw strength while animals pulled simple platforms over dirt and stone. These sledges created friction, which slowed everything down and made long trips exhausting. A short journey could take most of the day. A long journey could take many days and require large groups of workers.

The idea of putting a rotating wheel underneath a load came from careful thought and problem-solving. Early builders already knew how to shape wood into smooth, round surfaces for potter's wheels. They understood balance and how to steady circular motion. When they applied these skills to transportation, everything changed. A sledge that once scraped along the ground now rolled forward with far less resistance.

Archaeology reveals some of the earliest wheels from both Mesopotamia and southeastern Europe, suggesting the invention may have occurred in different places around the same time. The oldest known wheel, found in the marshlands of Slovenia, dates back more than 5,000 years. It was made from well-selected wood and showed signs of expert craftsmanship. Meanwhile, ancient drawings and clay models in Mesopotamia show four-wheeled carts pulled by animals. These discoveries suggest that the wheel did not spread from a single inventor. Rather, different societies figured out similar solutions because they faced similar problems.

When the first carts rolled across the ground, they allowed more supplies to move faster. Farmers could bring grain to market with energy to spare. Potters could ship jars farther away to new customers. Families could travel with more belongings. Distance no longer felt like an enemy.

Solving the Great Wood Problem

Transforming a tree into a reliable wheel was not easy. Builders had to cut a disk round enough to roll smoothly, yet strong enough to hold weight. Early wheels were often made from a single piece of wood, carved carefully into shape. Cutting that circle required patience, steady hands, and awareness of where a tree might split or crack.

Axles, the rods connecting wheels, also posed challenges. If an axle fit too tightly in the center hole of the wheel, the wheel couldn't turn. If it was too loose, the wheel wobbled and could slide off. Early craftsmen learned to adjust and smooth both the axle and the hub of the wheel, sometimes adding animal fat to reduce friction. They improved the design through trial and error, guided by experience.

Over time, carts became longer-lasting and more efficient. Builders experimented with two-wheel and four-wheel forms and tested how much weight each cart could carry. They also learned to repair wheels often because wood swelled in rain and shrank in heat. Maintaining wheels became a crucial part of travel, just as filling fuel is today.

These early engineering problems may seem simple to us now, but solving them opened a door to movement that would change how civilization grew.

The Power of Animal Partnership

The wheel became truly useful only when paired with domesticated animals. Oxen, donkeys, and other working animals already helped farmers plow fields and carry heavy bundles. Once hitched to a cart, however, the power of an animal multiplied greatly.

A strong pair of oxen could pull a load that once required many people. This new power changed everything. Farms expanded, harvests increased, and trade routes grew longer. With more grain and more goods available, populations grew. Villages became small towns, and small towns evolved into early cities like Uruk and Eridu.

Animal power also pushed new experiments in transport design. Harnesses and yokes spread weight more evenly across the animal's body so it could pull heavier carts without injury. With each improvement, travel became easier, and more people could focus on skills other than farming. Specialized craftsmen and merchants began to appear more often in society.

The wheel and harness formed a partnership that shaped the future. Without each other, neither would have transformed the ancient world so powerfully.

Different Lands, Different Needs

Even though the wheel was extremely helpful in some areas, not every region found it useful. The land itself often decided. In dense forests where trees stood tightly packed, carts could not pass easily. In sharp mountain regions, narrow paths made wheels less practical. Marshes and thick mud slowed carts as much as sledges.

In the Americas, people developed wheeled toys, proving they understood how wheels worked. Yet they had no large domesticated animals capable of pulling heavy carts. Llamas could carry loads on their backs, but they were too small for wagons. So Indigenous societies built excellent footpaths and bridges instead, especially in the Andes where the Inca created long and interconnected networks.

These choices were not mistakes. They were practical responses to different environments. When cultures needed the wheel, they invented or adopted it. When they didn't, they used solutions that fit their landscape.

Building Roads and Expanding Cities

Once carts became common in places like Mesopotamia, roads grew important. Dirt paths widened into packed routes. Low walls or stones sometimes lined the sides to mark directions or prevent carts from slipping. As more goods traveled, markets grew busier. Traders could bring pottery, cloth, and food from miles away.

Cities expanded outward and upward because transportation made it easier to build. Timber and stone arrived from farther distances. Workers brought tools and crops into new urban areas. Accessibility shaped where cities formed, and many rose along riverbanks where boats and carts both traveled well.

Government and record-keeping also changed. When goods moved farther and faster, officials needed ways to track ownership, taxes, and trade agreements. Clay tablets with cuneiform writing became tools for organization. The wheel helped carry information as much as it carried products. Soon, the movement of carts became tied to the movement of ideas.

Lighter, Faster, Better

Once people understood how useful the wheel could be, they began to improve it. Solid wooden wheels were strong but heavy. A heavy wheel slowed carts on long trips. To solve that problem, some builders cut away wood to leave only an outer ring supported by spokes.

Spoked wheels were lighter and faster. They allowed for greater speed when pulled by animals that could run, such as horses. Though solid wheels remained common for heavy loads, spoked wheels gained importance for transport that needed speed. Their use would eventually help create fast chariots, tools for communication, trade, and battlefield shock.

Every improvement made the wheel more adaptable. Inventors found ways to repair wheels faster, fit stronger axles, and shape hubs that held everything together more securely. With each success, the wheel spread farther across continents.

Rotation Beyond Travel

Once people saw how rotation could convert energy into motion, they began using wheels for more than carts. The same idea powered other tools. Waterwheels turned running streams into energy that could grind grain. Machines in workshops used wheels to wind ropes and lift heavy objects. Later, gears helped convert circular movement into different speeds and directions.

The wheel taught people how to make work easier through motion that did not require constant muscle force. It became part of construction, craft, and eventually timekeeping. In this way, the wheel supported the birth of engineering and the spread of mechanical thinking.

A simple circle became a source of power that shaped nearly every tool that followed.

A Turning Point in Civilization

The wheel's invention did not rely on a single genius. It came from many unnoticed workers who shaped wood, trained animals, hauled goods, and fixed broken carts in the dust along early roads. With every chip of the axe and smooth press of the file, ordinary people improved something extraordinary.

The wheel shortened distances, strengthened communities, and encouraged exploration. It linked farms to cities, cities to kingdoms, and eventually kingdoms to empires. It helped create trade routes, new economies, and ways of life that depended on movement and exchange.

Back in that early workshop, the potter finished shaping a clay bowl and raised it into the light to check its form. He set it aside and placed a fresh lump of clay on the spinning wheel. He could not know that his simple tool was the model for wagons, chariots, and countless machines to come. He only knew that the circle beneath his hands turned smoothly and helped him shape the future, one rotation at a time. The wheel kept spinning, and the world soon followed.

Chapter 2
Thunder on the Plains – Chariots and the Rise of Mobile Warfare

Dust clung to the air as dawn stretched across a wide open plain. Warhorses stamped the ground, snorting clouds of steam into the cool morning. Warriors balanced on lightweight platforms with narrow wheels, gripping the railings firmly as the signal to charge neared. Their chariots pointed toward the enemy line like arrows ready to fly. When the command finally rang out, wheels spun into action, hooves hammered the packed earth, and a storm of dust swallowed the plain. Spectators watching from a distance felt both awe and fear. Nothing on a battlefield moved faster than a chariot. Nothing struck more suddenly. Warfare had changed. It was now controlled by those who could command speed.

Chariots were not simply carts with horses. They were precision-built vehicles that introduced mobility, shock, and strategy into battle. Their rise marked a turning point in ancient power, when speed became a weapon all its own.

From Load-Bearers to Weapons of Speed

The earliest wheels were heavy and slow. They helped farmers and traders move goods, not soldiers. Solid wooden wheels made carts strong enough to carry weight, but they were far too heavy for fast movement. Over time, inventors realized that removing wood from the center of a wheel did not weaken the rim if carefully designed spokes supported it. These spoked wheels were lighter and spun faster.

With faster wheels came a new idea: what if carts could be used for war? Builders replaced solid frames with lighter ones made from carefully shaped wood. They experimented with axle length, wheel placement, and balance. The best designs kept weight toward the back so the chariot did not tilt forward when the horses galloped. The platform itself was kept small and open, just large enough for a driver and a fighter. Too much weight meant too little speed.

Harness design also needed to change. Older harnesses pushed into a horse's windpipe, limiting breathing. When warriors discovered harnesses that pressed instead on the chest and shoulders, horses pulled more safely and powerfully. These changes created a true military vehicle, one that carried warriors swiftly across the earth with force and control.

Once the chariot was ready, military leaders immediately understood what it could do. A soldier who once ran toward the enemy on foot now arrived like a storm.

Egypt: Chariots of Royal Power

In ancient Egypt, chariots became symbols of royal might. Wealthy nobles trained horses, polished bright wheels, and installed leather coverings and fine paint. Pharaohs used chariots in both warfare and ceremony. They rode them into battles to direct troops and to display dominance over enemies. Relief carvings show pharaohs aiming bows from speeding chariots, striking enemy forces with both skill and performance.

Chariot warriors held elite status. They practiced how to turn sharply without tipping and how to fire arrows while racing forward. Their horses were carefully bred and trained to obey commands even amid chaos. The chariot corps represented strength and control. They proved that whoever mastered speed could also master the land around them.

While artwork and inscriptions celebrate great victories, historians must be careful. Egyptian art often exaggerated the pharaoh's role in battle. A carving might show one ruler defeating whole armies alone. Yet real battles required teamwork, skill, and logistics, not just royal bravery. Even so, the pride Egypt placed in chariots reveals how much this vehicle changed warfare.

The Hittites: Power Through Heavy Impact

While Egyptians favored speed and archery, the Hittites of Anatolia turned to heavier chariots for shock combat. Their vehicles carried three warriors instead of two: a driver, a shield-bearer, and a spearman. This gave them more direct striking power. A line of charging Hittite chariots acted like a battering ram that could break enemy formations.

The Battle of Kadesh, fought around 1274 BCE between Egypt and the Hittite Empire, is one of the most famous chariot battles in history. Both sides used hundreds of chariots. Egyptian accounts describe dramatic triumph, while Hittite evidence points toward a stalemate. Propaganda aside, the battle showed how chariots dominated Bronze Age warfare. Whoever controlled faster movement controlled the fight.

To support such a force, the Hittites needed skilled chariot makers, trained horses, and constant access to wood and metal. Chariots were expensive and required regular upkeep. Only strong kingdoms could maintain them. Thus, military success and political strength were tied to technological ability.

Mycenaean and Aegean Uses: More Status Than Strategy?

In the Aegean world, especially among the Mycenaeans, chariots appear frequently in art and in the tablets from Pylos and Knossos. Yet evidence suggests they may have been used more for transport and ceremony than heavy combat. The rocky landscape made chariot maneuvering difficult, and infantry remained the backbone of Mycenaean armies.

Still, owning a chariot maintained the status of warriors known as the wanax class. For them, the vehicle showed wealth, noble heritage, and readiness for war, whether or not the battlefield allowed for wheels.

Here, the chariot's power lay more in image than in strategy, but image itself could strengthen a ruler's authority.

China: Chariots of Command

Chariots also played a major role in early Chinese warfare, especially during the Shang Dynasty. Royal graves contain chariots buried with important leaders, suggesting strong symbolic and military value. Chinese chariots were often used as mobile command platforms where generals oversaw infantry and directed formations.

Horses were highly valued, and great effort went into maintaining stables and training drivers. The chariot helped establish social ranks in Chinese armies. Those who rode in them belonged to noble families and expected loyalty from the foot soldiers below.

Bronze weapons, such as axe heads and spear tips, paired well with the speed of a chariot charge. The combination made Shang armies feared and respected.

Technology and Class: A Warrior Elite

Chariots did not belong to common soldiers. It took wealth to own horses, wheels, and the craftspeople required to maintain them. Skilled drivers and fighters needed long training and fine weapons. Supporting them required special stables, workshops, and hay fields.

The result was a small, powerful class of warriors who held both military and political influence. They were the fast-moving enforcers of a king's will, a reminder that speed did not just change battles, but societies.

A ruler with chariots could show strength at the edges of their territory more easily. Tribute could be collected faster. Revolts could be put down before they grew. Chariots made leadership mobile.

A Weapon With Limits

For all their strengths, chariots also had weaknesses. They needed open, flat ground. Rocks, steep slopes, and thick forests could stop them entirely. If a chariot wheel broke in battle, the warriors became easy targets. Poor weather made mud that trapped wheels and slowed horses.

Enemies learned to adapt. Some armies used terrain to their advantage, hiding behind hills or building traps. Before long, riders found a new way to use the horse itself: directly in battle.

Cavalry, soldiers fighting on horseback, could go where chariots could not. They were more flexible, faster to train, and easier to replace when lost. By the time iron weapons spread widely, cavalry had begun to overshadow chariots. The era of the roaring wheel was slowing down.

A Thunderous Legacy

Although eventually replaced by mounted warriors, chariots left a lasting mark. They taught societies how to train horses for speed and obedience. They pushed advances in metal tools, wheel design, and bow construction. They gave rise to elite military classes and changed the structure of ancient governments.

More than anything, chariots proved that movement could be a form of power. For a few centuries, the wheel and the horse working together shaped the borders of kingdoms and the outcome of wars. Armies that once marched slowly on foot now struck like lightning across the plains.

The sun set again over battlefields where wheels once cut narrow lines through the dirt. Chariots no longer thunder across those plains, but the idea they carried—that control of speed is control of strength—remains a permanent lesson in the story of war.

Chapter 3
Roads of Stone – The Royal Road and Persian Imperial Control

A lone courier urged his horse forward, the animal's hooves pounding against the packed earth of the Royal Road. The sun beat down on the long stretch of land rolling out ahead of him, disappearing over the horizon. He had already changed horses twice that day, swapping a tired animal for a rested one at a relay station guarded by soldiers loyal to the Great King. Dust rose behind him in a long plume, marking both his speed and the urgency of his mission. In his satchel, a sealed message held orders that would shape what happened hundreds of miles away. There was no time to slow. The unity of a vast empire depended on the speed of his movement.

This was the Achaemenid Empire under Darius I, a collection of peoples, languages, and lands spanning thousands of miles. Without strong communication, such a territory would break apart. The Royal Road stood as the empire's backbone, binding distant regions to the king's authority. The courier galloped onward, confident that the road and its system of support would carry him the entire way.

An Empire Built on Connection

When Darius I took the throne in the late sixth century BCE, he inherited a large kingdom stretching from the Indus River in the east to the Aegean Sea in the west. But size alone did not make a state powerful. Control required the ability to move information, goods, and soldiers quickly. Rebellions could rise in weeks, and armies marching slowly across rugged land might arrive too late to stop them.

Darius organized the empire into provinces called satrapies. Each had a satrap, or governor, responsible for collecting taxes and enforcing royal laws. Yet even the most loyal administrator needed guidance from the capital. Without reliable communication, local rulers might drift toward independence or form alliances that threatened royal authority. The king needed a stable, fast network linking every satrapy back to him.

The Royal Road was the answer. Older paths already crossed parts of the empire, used by traders and local officials. Darius did not build a road from nothing. Instead, he strengthened, straightened, widened, and guarded what was already there, transforming scattered routes into a cohesive imperial highway.

In doing so, he gave the empire structure, a physical shape that matched its political ambitions.

Engineering for Efficiency

The Royal Road stretched roughly 1,600 miles between the ancient cities of Susa in Persia and Sardis in Anatolia. Its length crossed mountains, plains, and stretches of desert. To maintain such a route required careful engineering. Workers leveled roadbeds where possible and packed the surface with gravel or stone. In trouble spots where carts and horses slowed, repairs were frequent and expected.

Road markers helped travelers keep steady direction. Bridges and causeways allowed easier passage across rivers and marshy areas. Although much of the surface has eroded or disappeared over time, archaeological evidence in Turkey and Iran shows straight alignments and foundations matching descriptions from ancient texts. These remains help historians trace the route even today.

Herodotus, a Greek historian writing in the fifth century BCE, marveled at the road's speed and efficiency. He described post stations spaced along the entire route, offering fresh horses and guards. His account, though not perfect, gives insight into how the road impressed foreign observers who recognized its power.

Every stone, bridge, and outpost served the same purpose: to make movement easy and dependable in a kingdom too large for any ruler to govern without help.

Relay Stations and Rest Stops

One of the most remarkable features of the Royal Road was its posting system. Relay stations, called pirradazis, were placed at regular intervals. Here, couriers could switch horses quickly. Some stations also offered fresh food, secure lodging, medical care for both humans and animals, and clean water.

These stations were staffed by trusted workers who monitored who passed through. They verified identities and messages, ensuring that only authorized travel continued along restricted segments. Strangers or suspicious individuals might be turned back or detained.

This system made communication astonishingly fast. Herodotus claimed that royal messages could travel the road in seven days—an exaggeration, but one that still recognizes the achievement. In reality, urgent messages could cover the distance in less than two weeks, many times faster than common travelers.

Relay stations also helped merchants, diplomats, and sometimes ordinary travelers. Though not free, the services they provided gave people confidence that journeys could be completed safely, even across dangerous territory.

Trade and the Flow of Wealth

The Royal Road was not only used by the king's messengers. It soon became a major trade route connecting different parts of the empire. Caravans loaded with spices, grains, textiles, and metals traveled between cities. Markets flourished as imported goods reached new customers. Even distant regions began to depend on one another.

Tax collectors tracked goods moving along the road, charging fees that filled the imperial treasury. Standardized weights and measures were enforced at checkpoints so merchants could no longer cheat buyers in one city and change prices in another. Written records kept accounts of profit, ownership, and payment.

Because traders and goods could move quickly and reliably, life improved in many regions. Cities grew near busy road sections. Services such as stables, inns, and craft shops appeared to support travelers. The road became a living economic network, transmitting wealth across the empire with every passing cart and pack animal.

Security, Soldiers, and the Edge of Control

Maintaining a road of such importance required protection. Soldiers patrolled key segments, especially near borders where foreigners might enter or where bandits might hide. Guard towers watched road traffic. Checkpoints inspected travelers and verified documents. Law enforcement along the route displayed the king's authority to friend and foe alike.

The road also allowed rapid deployment of troops. When a distant region faced attack or revolt, cavalry and infantry could march using the road's firm surface. Food and supplies could follow in wagons. In earlier times, rulers might have spent months mustering help from the capital. Under Darius, soldiers could appear much faster, often before threats fully formed.

For the average person, this security allowed safer journeys. Merchants who once feared theft now trusted the road enough to carry valuable products. Even dignitaries from far-off lands traveled with less worry, knowing patrols were never far away.

Mobility was power, and the Persians held that power firmly—for a time.

Voices from the Past: Written and Physical Evidence

Herodotus described the Royal Road with famous admiration: "Neither snow nor rain nor heat nor darkness of night prevents these couriers from swiftly completing their appointed journey." His words praised the efficiency of Persian messengers. Although later adopted as a motto for modern postal services, the line originally celebrated Persian innovation.

Archaeologists have uncovered road foundations, settlement remains, and road markers supporting his narrative. Sections of the route in modern-day Turkey align with what Herodotus recorded more than 2,400 years ago. Clay tablets from Persian administrative archives include references to rations and supplies for messengers, further confirming the system.

Together, these sources show that the Royal Road was not just a legend. It was a real structure that held a real empire together.

Where Power Meets Vulnerability

The same qualities that strengthened Persia eventually exposed weaknesses. A road that moved royal messengers quickly could also speed enemy soldiers. When Alexander the Great marched east centuries later, he used many of the same routes Persia had built to dominate its own lands. He moved faster than resistance could form, taking advantage of roads designed for imperial efficiency.

The Royal Road, once a symbol of Persian unity and intelligence, became a path for the empire's downfall. It demonstrated a truth repeated throughout history: infrastructure can empower a nation or assist its conquerors, depending on who controls the route.

Mobility made Persia strong, but when mobility shifted into the hands of another, Persia was left vulnerable.

A Lasting Route Through Time

The Royal Road did not disappear when empires changed. Many sections continued in use, repurposed by later rulers who recognized their value. Trade still flowed. Travelers still passed from city to city. The idea of a central road system influenced later infrastructure in the Roman Empire and far beyond.

It served as a reminder that geography can be shaped by human effort. Roads allowed distant peoples to feel the presence of authority and opportunity. They tied many small worlds into one larger world.

On the long plain, the courier slowed his horse as he approached another station. He handed over his message and took a moment to drink. Soon he would ride again. The empire depended on him, and he depended on the road—a road of stone, strategy, and connection that made the world feel smaller and rulers feel nearer.

The road held the empire together. It carried orders, goods, and hope. And long after those first riders passed, its legacy continued beneath the footsteps of history.

Chapter 4
The World Paved in Empire – Rome's Highways to Conquest

The sun glimmered off freshly cut stones as Roman engineers surveyed a new military road stretching across a rolling countryside. Legionaries organized soil into neat mounds, dug trenches with practiced arms, and carried materials with rhythmic efficiency. A surveyor followed a tall wooden instrument called a groma, its crossed strings guiding him to create perfectly straight alignments. Hammer strikes echoed across the land like the beat of a drum. Voices called out commands with urgency and focus. This was not temporary work. The Romans were building for centuries, not seasons.

A Roman road did not merely lead somewhere. It meant something. It was a statement of Rome's order and reach, laid down one stone at a time by men who believed their empire would last forever.

Layers Designed for Ages

Roman engineers built with the belief that infrastructure should outlive its builders. A standard road began with a deep trench dug into natural soil. At the bottom lay the statumen, a foundation of large rocks placed to support weight and prevent sinking. Above it came the rudus, a strong mixture of gravel and lime that formed a stable platform. Then the nucleus, a layer of smaller gravel or sand, was packed to become nearly as hard as concrete. Finally, the top course—called the summum dorsum—featured large paving stones fitted so tightly that a blade of grass could barely slide between them.

Each layer had a purpose: strength below, smoothness above, and drainage in between. The entire system worked together like a machine underfoot. A slightly rounded center created a slope that allowed rainwater to run toward the ditches, keeping the road surface dry. In a world where weather destroyed roads as quickly as armies did, drainage meant survival.

Modern engineers still admire this layered design for its durability. The fact that sections of Roman roads remain intact after two thousand years proves their success. Rome did not build shortcuts. Rome built permanence.

Straight Lines Through Mountains and Marsh

Romans respected geometry. When planning a new route, they drew lines across maps and landscapes, aiming for the most direct path. Curves appeared only when geography left no other choice. The result was roads that continued straight for miles, even across rugged terrain that would have made other cultures give up.

Survey teams crushed obstacles rather than go around them. They cut through hills, pushed causeways over marshlands, and built retaining walls on steep slopes. Where rivers flowed deep, they built bridges of stone or timber, joining land masses that nature kept apart. In Italy, fragments of such bridges still stand, silent witnesses to Rome's determination to claim the land itself.

Tunnels cut through rock ridges to keep routes direct. In the Alps, Roman roads carved high into the mountains gave traders and soldiers pathways once considered unreachable. When travelers marched along these roads, they walked directly into the heart of Roman ambition.

Milestones as Messages of Power

Along the roads stood cylindrical stone markers called milestones. Each listed the distance to the next major destination and included an inscription honoring the emperor responsible for construction or repairs. They served as maps, signposts, and political announcements carved into rock.

At the center of Rome, in the Forum, rose the Milliarium Aureum, the Golden Milestone. It marked the symbolic point from which all major roads radiated outward. To Romans and foreigners alike, it reinforced a powerful idea: Rome was the heart of the world, the place where all journeys truly began.

When merchants or soldiers stood beside a milestone in a distant province, they were reminded that invisible threads tied them back to the capital. A milestone did not just tell distance. It told allegiance.

Roads that Carried Legions Like Thunder

Military needs drove Rome's roadbuilding. Armies had to move quickly to secure conquests, defend boundaries, or crush rebellions. A legion marching along paved roads could cover enormous distances, arriving well-fed and ready to fight. Supply wagons brought food, weapons, and medical gear with less risk of getting stuck or delayed.

Speed meant surprise. Rome's enemies could not gather strength when legions appeared sooner than expected. Roads also allowed governors to call for help without waiting months for reinforcements. Even faraway territories felt the weight of Roman presence. Power traveled on wheels and sandals.

As Rome expanded across Europe, North Africa, and the Near East, roads made those lands feel closer to one another. An empire built on conquest remained strong by movement.

Trade and the Flow of Everyday Life

When peace settled across the empire, roads shifted from military purpose to economic backbone. Traders loaded carts with pottery, metalwork, spices, olive oil, grain, wine, and countless other goods. Cities that once relied only on local supplies now tasted foods and saw crafts that came from hundreds of miles away. A villa owner in Gaul might decorate a home with glass imported from Syria. A soldier stationed in Britain might sip wine from vineyards near Rome.

Markets along roadways flourished. Farmers could sell surpluses more easily. Inns and taverns welcomed travelers. Blacksmiths, wheelwrights, and laborers found steady work repairing carts and maintaining equipment. Movement supported prosperity, and prosperity supported peace—at least for a while.

Governance Along the Highways

Roads also served the government. Officials and inspectors traveled quickly from city to city, allowing the empire to enforce policies and resolve disputes efficiently. Messages delivered by mounted couriers arrived in days instead of weeks. Bureaucrats monitored taxes, justice, and military readiness across continents.

In many provinces, Roman law, architecture, and language followed the roads. Latin inscriptions, Roman arches, and bathhouses appeared where none had existed. Roads promoted more than goods. They spread ideas, religion, and cultural identity. Travelers brought news and stories that shaped how people understood the empire—and their place within it.

A person living at the far edge of the empire could still feel part of Rome because roads made distance feel less distant.

The Major Routes That Defined an Empire

Among the many legendary roads, the Via Appia stood out. It began in Rome and stretched to Brundisium on the Adriatic coast, connecting the capital to eastern trade routes. Its paved surface of dark basalt stones offered a smooth ride for carts and chariots. The poet Statius once called it regina viarum, the queen of roads.

Other grand routes included the Via Aurelia, running along Italy's western coastline into Gaul, and the Via Claudia Augusta, which scaled Alpine passes to link Italy with Germanic territories. In Hispania, the Via Augusta traced a long north–south line that shaped settlement and commerce across the Iberian Peninsula.

Archaeologists studying these roads today find wheel ruts worn deep by generations of travelers. Landscape aerials reveal astonishing straight alignments, still visible after centuries of erosion. The endurance of these paths tells the story of an empire that refused to fade quietly.

Infrastructure That Survived Its Builders

Roman roads have outlived the empire that made them. Their remains continue to guide modern routes. Highways in Europe often follow the ancient alignments chosen long ago by Roman engineers who studied the land and made choices that remain practical today.

Fragments of old pavement sit preserved under glass in museums or embedded in the floors of modern cities. Each preserved section whispers about the traffic of the past: soldiers' boots, merchants' wheels, priests' sandals, messengers racing with scrolls sealed in wax.

Technological achievements come and go, but the best ones remain useful. Roman roads proved that good design can transcend time.

When Roads Made Rome Vulnerable

The roads that once strengthened Rome eventually exposed a weakness. As imperial unity faltered in later centuries, protection diminished. Invading forces—Visigoths, Vandals, and others—marched straight into the empire along the very paths designed for legions. Roads sped their advance.

What had once been a tool of control became a tool of invasion. This reversal showed that technology does not choose sides. Power belongs to whoever controls the road at that moment in history.

Even so, the roads remained. Rome fell, but the infrastructure did not.

Stones That Unified the Ancient World

A Roman road was not only a means of travel. It was a political message visible in every province: Rome holds this place. Rome connects this land. Rome brings order.

Roads turned conquered lands into a structured map of cities, farms, and trade hubs rather than scattered local powers. They created a network of life that made the empire feel like one entity, not many. They shrank distance, accelerated change, and delivered identity.

Standing beside those engineers once more, watching them align one last paving stone into a seamless surface, we can see why these roads mattered. They allowed Rome not just to take territory but to shape it—organization written into the very earth.

And long after the standards of the legions vanished from the world, the paths they carved remain beneath our feet. Roman highways unified an empire. And in some ways, they continue to unify the world.

Chapter 5
Between Sand and Summit – Andean Road Networks

Mist clung to the cliffs like a veil as handlers guided llama caravans along a narrow path carved from the mountainside. The animals stepped carefully, their hooves gripping uneven stone, while travelers carried woven baskets filled with dried corn, wool, and ceramic goods. Far below, clouds filled the void of steep valleys, giving the impression that the world ended at the edge of the road. Above, jagged peaks pierced the sky, catching the first rays of daylight. On these high-altitude routes, every breath was thin and every step deliberate. The Andes were harsh, but the people who lived there were determined to move, trade, and communicate across some of the most dramatic landscapes on Earth.

This was not a world where wheels rolled easily. It was a world mastered through knowledge of stone, elevation, and the sure-footed help of llamas. In this unlikely environment, ancient Andean builders created one of history's greatest road systems.

A Continental Backbone

The Andes run nearly the entire length of South America, forming a rugged spine of mountains. The civilizations that lived along these slopes could have remained disconnected, isolated by geography. Instead, they solved the landscape's challenges, transforming their mountains into a network that united millions of people.

The Inca Empire, which rose to power in the 1400s CE, inherited earlier pathways built by the Wari and other Andean cultures. Rather than start from nothing, the Inca improved existing routes, extending them into a system now called the Qhapaq Ñan, meaning "The Great Road." This massive roadway stretched more than 24,000 miles, linking deserts, rainforests, coastal plains, and towering ridges. Some sections were wider than city streets, while others narrowed to no more than a single footstep.

The Inca approached the landscape like engineers and mountaineers. They studied the land and adjusted every route to suit its conditions, showing remarkable adaptability in the face of extreme nature.

Climbing the Impossible

Constructing steady walkways across such terrain required creativity. In gentler areas, builders flattened paths and lined them with stone retaining walls to prevent erosion. Where the mountains rose sharply, they carved stairs directly into bedrock. Some stairways climbed thousands of steps, turning impossible slopes into manageable routes.

On open stretches, they laid paving stones and used gravel to prevent mud during rainy seasons. Where loose soil threatened stability, they built walls as support braces, turning entire mountainsides into carefully balanced terraces. Travelers might walk from coastal lowlands into the sky itself, climbing through ecosystems that changed every few miles, from tropical humidity to freezing winds.

These paths were not only functional. They were symbols of empire reaching into heaven's domain.

Bridges Above the Abyss

The Andes are cut open by deep river canyons, and reaching across them required innovation. In places where stone bridges were impractical, the Inca built suspension bridges from grass, reeds, and woody vines woven into thick cables. These bridges swayed in the wind and creaked beneath travelers' feet, but they held firm under loads of people and cargo.

Bridge maintenance was a community responsibility. Villages gathered for ceremonies several times a year to replace old ropes and renew the structure. Hand-woven strength combined with social unity to keep travelers safe.

The balance between durability and flexibility made these bridges marvels of prehistoric engineering. Even today, descendants of the Inca maintain the Q'eswachaka Bridge using the same techniques, proving the lasting success of their ancestors' knowledge.

The Lifeline of the Tampus

Distance alone would have limited the usefulness of the Andean road network, but the Inca anticipated this challenge. They created tampus—stone storehouses and rest stations—along major routes. Some tampus stored food like chuño, a freeze-dried potato that could last for years. Others housed blankets, clothing, weapons, and tools.

These stations were spaced so that travelers never had to go long without shelter. Officials, armies, labor teams, and messengers depended on tampus for survival. A soldier marching from Cusco to Ecuador might rely entirely on these supplies. They turned the empire into one connected, functioning body.

Management of these stations showed how organized the Inca government was. Everything worked with precision, timed with seasonal needs and reinforced through strong administration.

Swift Feet and Swift News: The Chasquis

Without horses, carts, or wheels, the Inca developed a communication system that rivaled the speed of mounted couriers elsewhere. Teams of highly trained runners called chasquis waited in relay positions near tampus. When a message arrived, one runner sprinted to the next station, handing off news that could travel hundreds of miles in only a few days.

Messages were stored in quipus, bundles of colored strings tied with knots that recorded numbers and sometimes narrative information. Chasquis learned how to read these encoded records. Some carried oral messages memorized through protocols they practiced daily.

The chasqui network allowed the Sapa Inca—the emperor—to know what was happening across the empire faster than most ancient rulers. Speed kept power intact.

Why Wheels Weren't Needed

The absence of wheeled transportation in the Andean world has led outsiders to underestimate its sophistication. But the truth is simple: wheels do not roll well on stairs, narrow cliffside paths, or steep switchbacks. Roads built across jagged peaks would trap carts rather than help them.

Instead, the Andean people relied on llamas for transport. Llamas could carry moderate loads, usually around 50–70 pounds, and navigate steep, narrow pathways far better than horses or oxen. They required less food and water and were perfectly suited for high-altitude conditions.

Rather than forcing a technology that did not match the environment, the Inca refined travel methods perfectly suited to their world. They proved that progress is not measured by universal inventions, but by how well a culture adapts to what it has.

Roads as Lines of Defense and Control

Roads were also tools of military strategy. When rebellions arose or borders required defense, soldiers could rush along major routes with remarkable speed. Armies did not need wagons for supplies, thanks to tampus stocked in advance. Troops arrived alert and ready to fight.

Road checkpoints ensured that movement was monitored and controlled. Local leaders knew that rebellion would not go unnoticed for long. Through these systems, roads became part of political strategy as much as logistical support.

Imperial power traveled along these paths with authority that could be felt long before soldiers arrived.

Trade, Tribute, and Exchange

Although the Inca economy did not use currency the way others did, the movement of resources defined their wealth. Communities contributed tribute through labor, goods, or specialized skills. Alpaca wool, maize, coca leaves, copper tools, and finely woven textiles traveled along these roads, ensuring that each region benefitted from what others produced.

Shared goods brought cultural exchange as well. Stories, music, ceremonies, and languages blended across Andean society. The Qhapaq Ñan supported not just physical unity, but social transformation.

People living in distant valleys understood that the empire included them—that they were part of something vast.

Archaeology on the Edge of the Clouds

Much of this road system remains visible. Archaeologists hiking narrow ridges still walk on carved stones set in place more than 500 years ago. Satellite imagery has traced forgotten paths hidden beneath vegetation. Excavations near tampus and bridges reveal engineering choices that solved problems we struggle with even today.

Some roads retain their original drainage channels and retaining walls. Rope fragments from ancient bridges have survived due to dry Andean climates. Pottery near waystations helps identify trade patterns, proving that the roads carried not just armies, but daily life.

UNESCO recognizes the Qhapaq Ñan as a World Heritage Site, honoring its scale and ingenuity.

Empire in Every Direction

From desert to jungle, from coastal plains to glacial peaks, these roads carried the Inca message: unity through movement. The network tied together more than ten million people in a short time, demonstrating mastery over geography that would have overwhelmed other civilizations.

If you were to walk these routes today, you would feel the same solid stone underfoot that messengers once raced across. You would see mountain walls that still challenge travelers. And you would find tampus still clinging to cliffs, reminders that the empire planned for journeys that never end.

A Civilization Connected by Footsteps

The Andean roads remind us that greatness does not always roll on wheels or depend on smooth plains. Instead, it can cling to mountain faces and rise above clouds. It can test balance, courage, and resilience.

The Inca turned the Andes into highways of empire, allowing communication, government, and economy to function across extreme elevations. Their network was not a sign of luxury, but of profound organization and technological intelligence.

As the llama caravans continued along a high ridge, sunlight warmed the mist enough for views to reveal themselves. Terraces hugged mountainsides. Villages appeared in unexpected places. These pathways helped humanity conquer height itself.

Their legacy remains—not as ruins, but as living lines that still shape travel and identity in the Andes today.

Chapter 6
The Silk That Stitched the World — Asia's Overland Trade Corridors

The desert stretched empty beneath the rising sun, and the air shimmered with soft light as a caravan began its morning journey. Dark camel silhouettes moved steadily against the golden horizon, each animal carrying bundles of silk wrapped to protect them from sand and heat. Their broad feet sank quietly into the dunes, leaving faint impressions that the next windstorm would erase. The merchants guiding them were wrapped in long robes for protection. Ahead lay mountain passes, hidden oases, and distant cities where languages changed with every horizon crossed. The desert road seemed endless, yet the promise of profit and discovery urged the caravan onward.

This scene unfolded countless times across the network we now call the Silk Road. It was not a single straight path, but a large set of interconnected routes stretching across Asia. These pathways carried goods, peoples, and ideas between China and the Mediterranean. For more than a thousand years, they stitched together distant cultures into a dynamic world of exchange.

A Web of Routes, Not a Single Road

When people speak of the Silk Road, they often imagine a lone, unbroken highway from Xi'an to Rome. In reality, the routes branching across Eurasia were varied and shifting. Caravans traveled north of the Taklamakan Desert, passing through oasis towns like Turpan and Kashgar. Others moved south through mountain valleys where rivers cut narrow corridors. Some veered into India before reaching Persia. Still others traveled across the steppes into Central Asia and toward the Black Sea.

Which route a merchant chose depended on season, risk, and politics. If conflict made the northern path dangerous, caravans moved south. If drought struck or bandits controlled a mountain pass, trade shifted elsewhere. The Silk Road was a living system that responded to the realities of climate, economics, and war.

Because of this adaptability, even when empires rose and fell, trade seldom stopped. It simply found different paths.

Camels: Partners of the Desert

The Silk Road became possible in large part because of camels. The Bactrian camel, native to Central Asia, was especially valuable. With two humps and thick fur, it could carry heavy loads while enduring freezing nights in the mountains and scorching daytime heat in the desert. Its broad feet worked like natural snowshoes on sand, keeping it from sinking.

Merchants developed saddles designed to balance weight across the animal's back and panniers to protect fragile goods like silk and porcelain. Water bags, leather harnesses, and feed sacks completed the pack system that kept caravans moving.

At night, the camels rested in caravanserais, large roadside inns where traders could sleep safely, feed their animals, and share news. These waystations became places where tales traveled as quickly as goods.

Valuable Silk and Other Treasures

Silk gave the trade network its famous name, and its value was unmatched. Chinese weavers guarded the methods of silk production as a state secret. Silk's shimmering threads were lightweight, strong, and cool to the touch. When Roman elites first saw silk garments, they reportedly stared in awe, believing the fine fabric was made of magic.

But far more goods moved along these corridors. Spices like cinnamon and pepper flavored dishes across continents. Paper traveled west from China, transforming education and governance in places that adopted it. Glassware made in Roman workshops moved east, admired for beauty and shape. Precious stones, metals, carpets, medicines, and dyes changed hands often. Some goods traveled only part of the route before being sold again and again in a chain of profit.

Ideas traveled as well. Astronomical knowledge crossed borders, leading to improvements in mathematics and navigation. Information about medicine, farming, and craftsmanship spread with the caravans. Every trade carried something new.

Knowledge, Faith, and Culture on the Move

The Silk Road carried beliefs that reshaped ancient societies. Buddhism traveled from India into Central Asia and China, with monasteries rising along caravan routes. Stone statues found in places like Dunhuang reveal a blend of artistic influences, some featuring Greek-style robes and facial features that show how cultures mixed through contact.

Islam later flowed eastward across Persian and Central Asian trade towns, carried by missionaries and merchants. Christianity also moved along Silk Road paths, leaving carved crosses in places far from the Roman world.

Beyond religion, culture blended in everyday life. Foods changed. Music borrowed rhythms and instruments from distant places. Clothing styles incorporated new fabrics and techniques. Languages absorbed foreign words that remained in use long after their origins were forgotten. The Silk Road enriched the imagination of everyone connected to it.

Alliances and Agreements of Trade

Trade routes needed safety, and many rulers recognized that protecting caravans brought revenue into their cities. Empires formed agreements to guarantee safe passage and to tax goods passing through their borders. Market inspectors monitored transactions to enforce fairness and duty payments.

When a powerful empire dominated large segments of the route, trade often flourished. The Han Dynasty secured eastern sections. The Parthians controlled vital territory in Persia and acted as intermediaries between East and West. Later, the Sogdians, skilled merchants from Central Asia, became expert middlemen who built their fortunes on Silk Road commerce.

Yet treaties did not always hold. Rival states sometimes targeted caravans to weaken neighbors. Raiders sought profit from unguarded goods. Even successful traders planned for danger, traveling in groups strong enough to defend themselves. A peaceful road was rare, and always temporary.

Mountains, Deserts, and Shifting Borders

Geography shaped the Silk Road's possibilities and dangers. The high passes of the Pamir Mountains pushed travelers into thin air where breathing became difficult. Roads turned into narrow ledges where one misstep meant disaster. Snow blocked routes unpredictably, and avalanches claimed both lives and cargo.

The Taklamakan Desert posed a different threat. Leaving the path meant likely disappearance in a sea of shifting sand. Sandstorms blinded travelers, and heat drained energy quickly. Guides relied on knowledge passed through generations—where to find water, how to spot signs of an oasis, and when to stop and wait out danger.

Borders added human challenges to natural ones. A route safe one year might close the next because of war or a new ruler's demands. Caravans often changed direction to avoid conflict. Success required flexibility and a strong network of connections.

Cities That Thrived on Exchange

Along popular sections of the trade routes rose wealthy cities where cultures met. Samarkand and Bukhara glittered with markets offering goods from many lands. Kashgar became a vital crossroads, where caravans chose northern or southern paths around the Taklamakan. Palmyra prospered at the edge of the Roman Empire, its merchants serving as translators and brokers between different worlds.

These cities showcased architecture shaped by many influences. Temples housed more than one religion. Traders from China, Persia, India, Arabia, and Rome met in shared markets. Festivals and gatherings introduced unfamiliar customs to local populations.

Cultural identity in these cities did not belong to one people. It became a tapestry woven from many threads.

Messages That Traveled Faster Than Goods

Not everything carried along the Silk Road had weight. Stories of distant lands stirred curiosity. New knowledge about astronomy or engineering spread through conversation. Diplomatic messages moved as part of official caravans, shaping politics between empires. News about threats or opportunities traveled quickly because traders crossed borders often.

Even myths moved along these routes. Ideas from far-off lands gained their first audience through caravans, like tales of griffins guarding gold in distant mountains or fables of magical herbs growing at the world's edge. Information became a commodity exchanged by travelers.

War and Decline Along the Corridor

As centuries passed, power shifted. The Roman Empire weakened, and political turmoil disrupted western routes. Later Mongol expansion created a brief surge of renewed safety under the Pax Mongolica, when merchants could travel long distances under Mongol protection. But even that era faded.

Increasing sea trade changed world economics. Ships sailing across the Indian Ocean delivered goods faster, more cheaply, and with fewer dangers. The Silk

Road never disappeared, but its significance diminished as maritime exploration grew. The quieting of the caravans marked the end of a grand chapter in world history.

A Network That Connected Humanity

Even after trade slowed, the legacy of the Silk Road endured. Languages retained borrowed words. Art continued to reflect blended styles. Religions spread far beyond their original homes. Farming techniques and medical knowledge influenced generations living far from where they began.

The network proved that humans were not isolated, but capable of linking across vast distances through trade, curiosity, and shared needs.

As the caravan in the desert continued west, the sun rose fully and golden light bathed the path ahead. Every footstep pressed into the sand added to the long history of exchange. Goods would change hands. Ideas would cross borders. Cultures would transform. The Silk Road did more than move silk. It stitched the world together.

Chapter 7
The River Kings – Egypt and the Mastery of the Nile

The wooden vessel glided silently along the calm waters of the Nile. Its triangular sail filled with the steady north wind as the current pushed the boat from south to north, creating a natural rhythm that guided the journey forward. The water rippled gently against the hull, and the scent of fresh reeds drifted through the air. On both sides of the river, lush green fields stretched outward like a vibrant ribbon of life. Farmers bent over their crops while their families tended livestock along the banks. A short distance beyond the fertile valley floor, the earth changed abruptly to sand. The harsh desert blazed under the sun, empty and silent. Here, the contrast between life and death could be measured by a single step away from the water's edge.

From the deck of the boat, the scene looked timeless. Fishermen in smaller reed crafts cast their nets, hoping for a good catch. Children waved from the riverbank, their laughter carried over the water. The Nile was full of movement and sound. At any moment, a traveler could pass a cargo barge, a nobleman's pleasure boat, or a priestly procession heading toward a temple dock. The river was not just a route used by a few. It was a central part of everyone's life, from common workers to divine kings.

The Gift of the River

Egyptians knew that without the Nile, their land would be nothing but desert. They depended on the river's annual flood to bring fresh soil and guarantee survival. Far away in the African highlands, seasonal rains caused the Nile to rise each year, spilling onto the valley floor. When the water receded, rich black silt covered the farmland, making the soil fertile and ready for planting.

This predictable cycle shaped the Egyptian calendar. Farmers prepared fields as the waters pulled back, planted crops in the growing season, and harvested before the next flood began. Festivals and prayers honored the gods who controlled the waters. Egyptians thanked Hapi, the god of the flood, for life itself.

The rhythm of the flood also supported transportation. When the water rose, boats could reach areas far from the main river. Canals allowed ships to travel directly to farmlands, quarries, or temples. Egypt's wealth and strength were built on this dependable pattern year after year.

A Built-In System of Movement

Nature gave Egypt a rare advantage. The current flowed north. The wind blew steadily south. If travelers wanted to go north, they let the river carry them. If they wanted to go south, they raised their sails to catch the wind. By adjusting the sail or the rudder, a single boat could reach any point on the Nile and return without needing special roads or immense effort.

This natural two-way highway allowed goods, messages, and people to move quickly throughout the kingdom. Government officials could reach outlying villages. Soldiers could respond swiftly to trouble. Merchants could bring pottery, linen, jewelry, and other goods to markets hundreds of miles away.

Other civilizations struggled to build roads or cross mountains. Egypt's transportation system already existed. It flowed through the center of the country.

Boats from Reeds and Wood

At first, Egyptians sailed in small reed boats. Papyrus plants grew in thick clusters along the riverbanks. Stalks could be cut, bundled, and shaped into light, buoyant vessels. Hunters used reed boats to stalk water birds. Fishermen used them to set nets in shallow pools. Even today, similar craft still glide through parts of the Nile's marshes.

But as cities rose and trade increased, reed boats became too small and weak for the heavy loads needed to sustain the growing kingdom. Egyptians needed larger, stronger ships. They turned to wood.

Egypt lacked large forests, so shipbuilders imported cedar from Lebanon, rowing and sailing along coastlines to bring back timber. Egypt's rulers valued this wood so highly that they recorded cedar deliveries as royal achievements on temple walls.

Shipwrights built long wooden boats with curved hulls that could carry workers, tools, and cargo across great distances. They used mortise-and-tenon joints—wooden pegs and carved slots that locked planks tightly together. Ropes tied the beams in place, allowing the ships to flex just enough to handle waves and wind. Strong masts lifted linen sails. Oars lined the sides for additional control. These innovations turned the Nile into a network of commerce and construction.

Carrying Stones for Kings

Some of the largest boats ever built in ancient times traveled the Nile to construct Egypt's monumental buildings. When pharaohs ordered new temples or pyramids, workers journeyed south to quarries to cut huge stone blocks. Granite came from Aswan. Limestone came from cliffs near Cairo. Giant obelisks carved from single pieces of granite weighed many tons.

These stones were too heavy for overland transport. The river became the only realistic pathway. Barges carried block after block northward. When they arrived near the construction site, workers dragged the stones to their final locations.

The Nile's role in pyramid building was essential. Without river transport, even the most skilled architects and builders could not have assembled the great pyramids, temples, and tombs that remain today.

Feeding Cities and Armies

Egypt's farms produced more grain than people needed. The surplus supported soldiers, workers, priests, and government offices. Ships full of grain traveled daily from farms to state granaries. From there, grain could be sent anywhere in the kingdom. It even served as currency, used to reward workers and fund large projects.

Because food could be shipped from plentiful regions to areas with smaller harvests, the kingdom remained stable even in difficult years. This security helped Egypt grow stronger. A well-fed population could focus on tasks beyond basic survival, leading to developments in culture, trade, and science.

Grain shipments were so important that protective structures were built near docks to guard against theft. The prosperity of cities like Thebes depended on the constant flow of food along the Nile.

Ceremony and Sacred Journeys

Religion played a central role in Egyptian life, and the Nile was part of that sacred world. Egyptians believed gods traveled by boat. Ra sailed across the sky by day and through the underworld at night. Osiris ruled over life, death, and rebirth—just as the river flooded, receded, and returned each year.

Temples housed ceremonial boats decorated with gold and jewels. During major festivals, priests carried statues of gods to river barges. Crowds gathered along the banks to celebrate as boats drifted past, surrounded by music and incense. Such events reminded people that the pharaoh and the gods controlled the river and the land.

Royal processions also used the river. Pharaohs traveled in brightly painted vessels, attended by servants and soldiers. Seeing the king sail by was a powerful message: the living god on earth moves wherever he wishes, and the kingdom follows.

Even after death, the journey continued. Large boats were buried with pharaohs to ensure safe passage into the afterlife. These vessels symbolized eternal travel through cosmic waters.

Life at the River's Edge

Nearly every Egyptian settlement grew near the river. Homes lined the banks. Docks bustled with activity each morning. Markets offered fish, bread, pottery, and textiles brought in by boat. Craftsmen worked nearby so goods could be loaded easily onto waiting ships.

Children learned to swim early. They watched their parents navigate boats and mend fishing lines. Stories, songs, and traditions centered around the water and its importance.

The river gave community to those who lived along it. Families visited relatives by boat. Farmers met traders in distant towns. Weddings and festivals drew crowds from both sides of the Nile, who rowed or sailed to join the celebration. Egyptians did not simply live beside a river. They lived through it.

Art and Archaeology: The River's Record

Much of what modern researchers know about river life comes from tomb art. Paintings and carvings show all kinds of vessels: cargo barges moving stone, royal ships with decorated cabins, and small hunting boats in papyrus marshes. These images preserve moments of everyday activity and ceremonial grandeur.

Excavations support the art. Archaeologists have uncovered anchors, paddles, and shipyard remains. Near the pyramids, they found full wooden boats buried in pits. One spectacular find near the Great Pyramid—the Khufu ship—was assembled with such precision that modern engineers studied it to understand ancient shipbuilding.

These discoveries reveal a civilization that advanced transportation far beyond what many had imagined.

A Kingdom Connected

The Nile acted as the backbone of Egypt's government. Messengers used boats to deliver orders and collect reports. Soldiers traveled along the river to protect borders. Tax officials visited farms, ensuring the pharaoh received his share of the harvest.

This efficient system allowed Egypt to remain unified across long distances and through many generations of rulers. The river prevented separation and rebellion by keeping distant regions closely tied to the center of power. As long as the Nile flowed, the kingdom remained whole.

The River's Role in Trade

The Nile did not serve Egypt alone. Ships sailed north to the Mediterranean, where Egyptian traders exchanged goods with neighbors. They brought gold, linen, papyrus, and grain. They returned with timber, copper, wine, and oils. Trade strengthened the economy and introduced new goods into Egyptian homes.

Foreign ships also sailed up the Nile to visit Egyptian ports. This movement added diversity to markets and ideas. Egyptians took pride in their role as a vital link between Africa, the Mediterranean, and the Near East. The river made Egypt a crossroads of commerce and culture.

Timeless Waters

As the boat continued drifting forward, the desert sun lit the horizon with gold. The lush farmland remained green and vibrant, nourished by the river's generosity. The Nile had been flowing long before Egypt became a kingdom, and it would continue long after the lives of the passengers aboard the vessel had passed into memory.

The river carried stones for temples, grain for towns, and news for leaders. It carried worship and celebration. It carried hope and the promise of renewal.

The Nile was not merely a route through the landscape. It created the landscape. It shaped the society that depended upon it. The river gave Egypt the power to grow, the mobility to unite, and the resources to build a civilization that still amazes the world.

Where the Nile moved, Egypt moved. Where it prospered, Egypt prospered. The river was Egypt's first tool, its greatest resource, and the unbreakable bond that turned a narrow valley into the realm of the River Kings.

Chapter 8
Riding the Monsoons – Indian Ocean Navigation

The dhow cut through the waves beneath a glittering night sky. Warm winds caught the curved lateen sail and pushed the vessel across the open sea. Sailors adjusted ropes with practiced hands, their bare feet braced against the wooden deck as the ship tilted with each rolling swell. The moon's reflection stretched into silver ripples that crawled toward the distant horizon. Ahead, a faint line of shoreline promised arrival by morning. Behind them, stars guided their route like old friends. The sea was dark and deep, yet these sailors felt confidence rather than fear. They understood these waters and winds. They trusted the monsoon.

This voyage was part of a centuries-long movement of people, goods, and ideas that connected continents. Long before Europeans began crossing oceans, sailors of the Indian Ocean—Arabs, Persans, Indians, East Africans, and Southeast Asians—

mastered the rhythms of seasonal winds. Their ships carried spices, ivory, gold, and pottery, but they also carried languages, religions, and stories. Across thousands of miles, the sea became a water-road just as significant as any land route on Earth.

Learning from the Wind

Sailors noticed early that winds on the Indian Ocean did not behave randomly. They shifted direction almost like clockwork each year. From roughly June to September, strong winds blew from the southwest toward India and Southeast Asia. Months later, the direction reversed. From November to February, the northeast monsoon pushed ships back toward Arabia and East Africa.

These patterns made long-distance travel possible. Sailors planned their departure based on the predictable change of seasons. A trader heading east from Oman would wait for the southwest monsoon to fill the sails and carry the dhow toward India. When ready to return home, the northeast monsoon would do the work.

Ancient geographers learned from experienced mariners. Greek and Roman writers recorded reports from merchants who traveled between Egypt and India. One such trader, Hippalus, is often credited with identifying the monsoon system in the first century CE. Many historians believe sailors already understood these winds intimately. Hippalus simply recorded knowledge that had existed for generations. Understanding the winds turned the ocean into a map of opportunity.

Sewn Ships and Lateen Sails

The vessels that traveled the monsoon waters were unique. The dhow, a type of sailing ship common in the western Indian Ocean, featured a triangular lateen sail. Instead of relying on square sails that needed wind from behind, this design allowed sailors to catch side winds and turn more easily. Monsoons did not always blow perfectly in the direction the sailors wanted to go. The lateen sail made navigation flexible.

The hulls of early dhows were stitched together with coconut fiber cords rather than nails. Planks were drilled with holes and sewn tightly, forming a strong, slightly flexible structure. This method helped the ship withstand rough seas and avoided the risk of metal corroding in salty waters. Layers of pitch sealed gaps to keep the ship watertight.

Crews gathered fresh water and food at each stop, storing dates, dried fish, and grains in tightly wrapped containers. The ships carried goods for trade: frankincense and myrrh from Arabia, gold and ivory from East Africa, spices from India, and porcelain and silk from Southeast Asia. Dhows were more than boats. They were floating markets that moved culture from one shore to another.

Ports Where the World Met

When ships arrived at port cities, markets came alive. Goods exchanged hands. Ideas changed minds. Words from different languages filled the air.

On the southwestern coast of India stood Muziris, a major port described in Roman records. Traders there bought black pepper, pearls, gemstones, and fine cotton textiles. Roman coins found in the region show how eagerly Mediterranean buyers wanted Indian luxuries. Ships from the Red Sea regularly stopped here before following the coastline farther east.

Across the sea, on the island of Zanzibar near modern-day Tanzania, merchants from Arabia and Africa met to trade ivory, ambergris, tortoiseshell, and enslaved people. Over time, an entirely new culture emerged as settlers intermarried with local communities. The Swahili language combined Bantu roots with many Arabic words, reflecting centuries of shared commerce.

Farther east, the Strait of Malacca linked the Indian Ocean to the South China Sea. Cities like Malacca and Srivijaya became multilingual trading centers. Elephant tusks might arrive from Africa the same day fine silk arrived from China. People ate spices grown thousands of miles away. Music and clothing blended into styles never seen before. Where ships gathered, civilizations blended.

Knowledge That Traveled Farther Than Ships

Sailors relied on information passed down carefully through families and maritime communities. A father might teach his son how to read the wind by watching cloud patterns or how to follow coastal shapes in low light. Navigators learned the stars, the behavior of currents, and the habits of birds that flew out to sea in the morning and back to land at dusk.

Some knowledge became written. Arab mathematicians refined methods for measuring latitude. Astronomers studied the sky to help sailors stay on course during long voyages. Sailors also carried mental maps of winds and tides. Being wrong could mean drifting into open ocean with no hope of return.

Cultural exchange also included religion. Islam spread along monsoon routes as traders settled in Indian and East African ports. Mosques appeared beside temples and shrines, showing how diverse beliefs lived side by side. These journeys changed the identity of entire regions. Trade was a teacher.

The Growth of Sea Empires

As trade expanded, kingdoms sought to control it. Wealth from maritime routes supported powerful rulers. Control of a major port meant influence over regional politics and access to goods that symbolized luxury and status.

Arabian cities like Aden in Yemen became key stops linking the Red Sea to distant markets. Indian rulers built protective fleets to guard harbors from pirates. In Southeast Asia, the Srivijaya Empire built its strength around maritime dominance. They controlled shipping lanes and taxed ships passing through their waters. Their capital became a hub where Southeast Asia met the wider world.

Trade with China also increased. The Chinese treasure fleets under Admiral Zheng He in the 1400s sailed into the Indian Ocean with large ships, carrying goods and diplomatic gifts. These voyages demonstrated China's reach and forged alliances with distant states. Trade shaped diplomacy. Ship routes reshaped power.

Lives Built on the Ocean

For sailors, the sea was both home and challenge. They risked storms and shipwrecks each voyage. Navigation errors could push them too far offshore. Pirates or rival forces sometimes attacked.

Yet seafarers gained status and wealth. A successful trader could return with spices worth more than gold. Cloves and nutmeg from Indonesia traveled thousands of miles because their flavor and scent were prized wherever they landed. The Indian Ocean economy thrived because each region offered something unique.

Families waited on docks for safe returns, celebrating journeys with storytelling, music, and feasting. Sailors brought more than goods back—they returned with news of distant lands, strange customs, and new food or clothing styles. They expanded their community's understanding of the world. The sea shaped lives as deeply as the land.

Currents, Coral, and Challenges

The monsoons made travel easier, but not safe. Coral reefs along East Africa's coast could tear open a hull if sailors misjudged their path. Shallow sandbars hid beneath calm water. Sudden storms threatened to overturn ships.

Sailors hugged coastlines when possible, stopping at known harbors for fresh supplies. They listened to the ocean and watched the sky for warning signs. Experience became a form of survival.

Winds sometimes shifted earlier or later than expected. Dhow captains calculated routes by the moon and stars, hoping weather matched the knowledge passed down to them. A single mistake could turn a voyage into disaster. Despite the risks, trade continued because rewards were high and knowledge made confidence possible.

Exchange That Changed the World

Goods that traveled across the Indian Ocean transformed societies. Pepper from India changed the flavor of meals throughout the Middle East and Europe. Chinese porcelain decorated homes in Africa. Arabian horses arrived in new lands. Coconut, bananas, and sugarcane spread across continents, changing diets and agriculture.

But the most significant trade might have been cultural. Languages blended. Clothing styles influenced one another. Music, storytelling, and art spread along the monsoon's path. Belief systems moved with traders—Hinduism and Buddhism traveled southeast toward Indonesia, while Islam traveled southwest to Africa and east into Malaysia. The sea acted as a bridge rather than a barrier. Culture did not stay confined within borders.

A Web of Connection

The Indian Ocean network mirrored the benefits of the Silk Road, except ships could carry heavier loads and travel greater distances more efficiently. Land routes could be blocked by war or rough terrain. The sea offered flexibility. Caravans might take months to cross desert sands. Dhows could reach distant ports in weeks.

As long as sailors understood the monsoon, opportunities stretched far beyond their own coasts. People who had never met before could exchange goods and ideas that shaped their futures.

The Indian Ocean became one of the earliest examples of globalized trade—not ruled by a single empire but built on cooperation, skill, and trust.

The Endless Voyage

The ship sailed onward beneath the moon. The warm monsoon wind pushed against the canvas sail, guiding the dhow toward its next port. Soon, the horizon would glow with the light of another coastal settlement—a place with different languages, different foods, and different customs, yet joined to this crew by the sea.

The sailors looked forward to the moment when their anchor would drop and they could step onto shore again. But they also knew their journey was not finished. After a season of trade, conversation, and waiting for the winds to change, they would climb back aboard and let the monsoon carry them home.

The Indian Ocean turned distant lands into neighbors. It linked continents into a shared world of ambition and curiosity. Riding the monsoons, sailors became the heartbeat of an ocean that brought prosperity and connection to millions, shaping the course of history with every turn of the wind.

Chapter 9
Masters of the Deep – Greek and Phoenician Seafaring

The trireme sliced forward through the waves, its bronze-sheathed ram glinting in the sun as it approached a fortified harbor. Dozens of rowers moved in perfect rhythm, their oars dipping into the water with synchronized precision. Above them, the captain watched the coastline, calculating distance and speed while giving clear commands. Seagulls circled overhead, drawn to the smell of fish and the shouts of dockworkers waiting on shore. The harbor was a flurry of activity. Sailors unloaded cargo from merchant ships while lookout towers scanned the horizon for possible threats. In this vibrant world of trade and conquest, the Mediterranean Sea was a battlefield, a marketplace, and a highway at the same time.

Both Greeks and Phoenicians mastered these waters. Their ships traveled farther and faster than those of earlier civilizations, and their knowledge of winds, stars, and coasts turned them into true rulers of the sea. Their advances in navigation and ship design shaped history through trade, colonization, and war. Some sailed for commerce. Others sailed for glory or survival. All depended on the power of the sea.

Phoenician Navigators of the Unknown

The Phoenicians excelled as sailors long before many other cultures built strong maritime traditions. Emerging from coastal cities such as Tyre, Sidon, and Byblos, they transformed limited farmland into a seafaring network of trade. Their strength was not measured in land but in routes, paths mastered across the sea.

Phoenician sailors developed advanced understanding of currents, winds, and star patterns. They used the North Star to maintain direction in darkness, a technique others later adopted. This navigation knowledge gave them confidence to sail far west into the Mediterranean and beyond.

Their reputation for exploration stretched to the edge of myth. Stories describe Phoenicians reaching distant shores such as the Canary Islands and perhaps even beyond the Strait of Gibraltar, which the Greeks called the Pillars of Heracles. Whether every tale is true or not, archaeological evidence of Phoenician goods found far from home shows they were daring mariners who pushed boundaries.

Ingenious Shipbuilders

Phoenicians also built ships that could withstand demanding voyages. Their merchant vessels carried large amounts of cargo securely over long distances. Shipwrights constructed curved hulls using mortise-and-tenon joints, strengthening the vessels against waves and storms. The bireme, with two rows of oars stacked vertically, increased speed and maneuverability.

Purple dye, crafted from sea snails, traveled in jars across the Mediterranean, becoming a symbol of wealth worn by nobles and royalty. Cedar wood from Lebanese forests served as both export cargo and essential material for ship construction. Glassware, ivory, and precious metals filled Phoenician ships that arrived at busy foreign markets.

Their ships were not only technical achievements. They were symbols of identity. They announced the arrival of experts in trade, diplomacy, and cultural influence.

Networks of Prosperity

Phoenician ports became centers for economic growth. At harbors from Cyprus to Carthage, goods were stored, taxed, and redistributed. The establishment of colonies expanded their reach even farther. Carthage, founded on the North African coast, later became a dominant power in its own right.

Merchants managed complex trade arrangements, exchanging raw materials and luxury goods between continents. Every successful voyage strengthened the

Phoenician network. Wealth flowed not from conquering land but from connecting people.

Through trade, the Phoenicians spread ideas, writing systems, and technologies across the Mediterranean world. Their alphabet influenced later Greek writing, which shaped the scripts used across Europe today. Their legacy extended long beyond their own era.

Greek Designs for War and Commerce

While the Phoenicians focused on trade, the Greeks adapted maritime technology for both economic and military strength. City-states such as Athens, Corinth, and Rhodes relied on ships to secure resources and assert influence.

Greek ship design reached new heights with the trireme. Sleek, fast, and effective in battle, it changed naval warfare. Three tiers of oars powered the vessel, and the bronze ram at the bow could splinter enemy hulls on impact. Naval combat now required precision, discipline, and speed. Rowers trained constantly to keep stroke patterns synchronized. A single mistake could throw off balance and change the outcome of a battle. Triremes turned the sea into a battlefield where strategy shaped the fate of nations.

The Maritime Empire of Athens

The Athenians used their navy to build an empire based on tribute and alliance. After the Persian invasions, Athens formed the Delian League, a military alliance designed to defend against future attacks. Over time, Athens used League contributions to strengthen its own fleet and influence. Members relied on Athenian ships for security, making sea power a tool of political control.

The sea connected Athens to grain markets in the Black Sea region. Merchant ships carried olive oil, wine, pottery, and metalwork. Trade enriched Athens and supported cultural growth in philosophy, theater, and architecture.

Without mastery of the sea, Athens would never have risen to such prominence. Navy and commerce fueled not only its economy but also its identity.

Harbors Built for Defense and Trade

Greek and Phoenician advances were not limited to ships. They built harbor structures to protect vessels and control maritime access. Defensive towers guarded bay entrances. Chains could stretch across harbor mouths to block enemy ships.

Lighthouses, such as the famous Pharos of Alexandria built later in the Hellenistic period, guided sailors safely to shore after dark. Earlier beacon fires served similar purposes in Greek and Phoenician ports. These landmarks reduced shipwrecks and strengthened maritime reliability.

Docks and warehouses organized cargo efficiently. Merchants rented space to store goods between voyages. Economic specialization grew. Sailors, dockworkers, and craftsmen formed communities focused on maritime trade. Ports became symbols of progress and prosperity.

Sea Battles That Shaped History

As naval technology advanced, so did conflict. Battles on water determined the outcomes of wars that shaped entire regions. In the early fifth century BCE, the Greeks faced an invasion by the Persian Empire. Although greatly outnumbered, the Greeks won crucial victories at sea.

At the Battle of Salamis in 480 BCE, Greek triremes used narrow straits to their advantage. Faster and more maneuverable than Persian vessels, they rammed and sank many enemy ships. The Persian fleet, unable to fight effectively in tight waters, retreated. This victory prevented Greece from falling under Persian rule, preserving Greek independence and culture.

Sea power influenced not only the defense of territory but also the direction of Western civilization. Ideas of philosophy, governance, and culture that developed in Greece might never have emerged without naval strength.

Competition That Drove Innovation

Rivalry between maritime powers fueled innovation. When one city-state or kingdom improved its ships or tactics, others adapted to catch up. Seafaring nations invested in shipyards, trained rowers, and studied navigation more closely.

New materials improved durability. Mathematics helped calculate distance and improve mapping. Political alliances often depended on naval support. Colonies demanded regular communication by sea, forcing improvements in logistics and ship construction.

The Mediterranean became a place of progress, where survival depended on speed, skill, and strategic thinking.

Trade That Shaped Culture

Every voyage carried more than merchandise. Greek pottery styles appeared in foreign markets, influencing local art. Phoenician metalwork introduced new styles and techniques. Religion also traveled by ship. Gods moved with merchants. Foreign deities sometimes found homes in new lands.

Language spread along trade routes. Coins helped standardize value and simplify distant exchanges. Food crossed borders. Grapevines and olive trees grew in places where they had never been seen before.

Travelers returned home with stories of unfamiliar customs and inventions. Knowledge expanded because ships transported curiosity across water. Maritime trade became a force of cultural transformation.

Marks on the Coastline

Archaeology has uncovered harbors, shipwrecks, and artifacts that reveal Greek and Phoenician seafaring. Bronze rams found underwater show the violence of naval battles. Amphorae recovered from wrecks reveal trade routes and cargo types. Remains of shipyards indicate organized construction work.

Ancient walls protected waterfronts. Stones still show grooves worn by mooring ropes. The remains of towers and dock foundations mark where busy ports once stood.

These traces prove that the sea was not a backdrop to history. It was the stage on which much of ancient life played out.

The Enduring Sea

The trireme neared the harbor entrance, rowers slowing their pace as the ship glided past stone towers. In the city above, fires flickered as merchants prepared for markets the next day. The sailors would soon step onto land, unload goods, and celebrate after days at sea. But soon the call of the waves would draw them back.

For the Greeks and Phoenicians, the sea was more than water. It was a path to survival, power, and discovery. It carried their hopes and dangers. The Mediterranean connected distant cultures into a shared world of rivalry and exchange.

Masters of the deep did not rule by land. They ruled by horizon. Their ships carried not only cargo but the momentum of history, turning the tides of war, shaping commerce, and proving that the sea could be guided by human skill and ambition.

Chapter 10
The Shipwright's Revolution – Innovations in Roman Naval Power

Hammers struck wood in steady rhythms that echoed across the harbor. Sparks flew as iron tools shaped thick planks into curved hull pieces. Carpenters wiped sweat from their brows while cranes lifted beams into place. The smell of pitch, sawdust, and saltwater filled the air. In the shade of unfinished ships, Roman officers studied blueprints and compared measurements, speaking urgently about new designs that would soon command the Mediterranean. Massive stone docks and wooden scaffolding lined the waterfront as workers turned raw lumber into instruments of power. Nothing about the scene suggested a people new to the sea. Yet Rome had once known little about shipbuilding. Its future now depended on learning fast.

Rome began as a land-focused civilization. Its strength came from disciplined legions who marched roads across Italy. Warfare was fought with shields locked tightly, not with sails catching wind. But dominance over land eventually required mastery of water. As Rome expanded, survival depended on controlling trade routes and protecting its coastlines. The sea was no longer a border. It was a bridge to distant territories and a battlefield where victory would decide the fate of empires.

From Land Warriors to Seafarers

Before the Punic Wars, Rome's navy was small and inexperienced. Carthage, a powerful maritime empire in North Africa, commanded fleets that ruled the waves with confidence. Carthaginian sailors had navigated coasts and storms for generations. Rome recognized that if it could not counter Carthage's strength at sea, everything it had conquered on land would remain vulnerable.

Rome acted quickly. When a Carthaginian quinquereme ran aground, Roman engineers seized the opportunity to study it. They examined the hull design, the placement of oars, the weight distribution, and the materials. They then built a replica and used it as a model to mass-produce ships. What Carthage had developed through centuries of experience, Rome copied in months.

But imitation alone would not secure victory. Rome needed an advantage that would neutralize Carthage's superior seamanship.

The Corvus and the Roman Solution to Naval Combat

The answer came in the form of the corvus. This Roman invention was a large boarding ramp with a metal spike at the end. Once a Roman ship neared an enemy vessel, the corvus dropped down, the spike driving into the opposing deck. This locked the ships together.

Suddenly, naval combat became land combat. This shift favored Rome's greatest strength: close fighting with swords and shields. Legionaries stormed across the corvus into enemy ships, turning skilled sailors into desperate defenders. Techniques like ramming and evasive rowing mattered less when battles were decided by hand-to-hand combat.

The impact of this device was dramatic. At the Battle of Mylae in 260 BCE, Rome—new to naval warfare—defeated a large Carthaginian fleet. The victory sent a message across the Mediterranean. Rome was learning the sea on its own terms.

Shipyards That Never Slept

To sustain its naval transformation, Rome established shipyards throughout the empire. These facilities combined speed with organization. Workers specialized in shaping keels, preparing planks, and attaching rigging. Timber arrived from forests across Italy and beyond. Ropes were woven from strong fibers. Iron was forged into anchors and fittings. The entire production line displayed Roman logistics.

Engineers standardized ship components so that construction could happen in multiple locations at once. Design plans ensured consistent quality. If one ship suffered damage, workers in a distant port could build or replace identical parts. Rome did not need a maritime tradition to excel. It needed structure, and organization had always been its greatest advantage.

Naval bases such as Ostia became essential military hubs. Fleets could shelter in protected waters, refit quickly, and launch with precision. As Rome expanded, new ports appeared in strategic regions, projecting Roman control across newly claimed seas.

A Fleet That Carried Armies

Roman ships were not only for battle. Transport vessels allowed Rome to move thousands of soldiers across seas in short periods. Instead of marching across rugged land, legions could board fleet transports and arrive in distant theaters ready for action.

During Julius Caesar's campaigns in Gaul, naval support allowed his troops to cross the English Channel and challenge tribes in Britain. Later, fleets secured territories across the Mediterranean, including Greece, Asia Minor, and North Africa. Ships carried cavalry horses, siege engines, medical supplies, and food needed to sustain long campaigns.

This rapid movement meant Rome could respond quickly to rebellion or invasion. The sea linked legionaries with support systems that made them nearly unstoppable.

Feeding the Heart of the Empire

Rome's population grew quickly, and feeding the capital became a major challenge. The Mediterranean provided the solution. Grain from Egypt and North Africa traveled in large transport ships to storage facilities near Rome. Without this regular flow of food, famine and unrest would have threatened stability.

The navy guarded these valuable shipments. Pirate raids could cause severe damage if allowed to strike freely. By keeping sea routes safe, Rome ensured that bread remained available. Prosperity at home depended on discipline at sea.

Wine, olive oil, spices, and other goods also traveled across the water. Commerce enriched port cities, fueling economic growth across the empire. Control of shipping lanes was not just military strategy. It was the foundation of everyday life.

The Mediterranean Becomes Mare Nostrum

After defeating Carthage, Rome expanded its naval dominance. The Mediterranean became known as Mare Nostrum, "Our Sea." This phrase reflected reality. Rome policed waters from Spain to Syria. Pirates hiding along rugged coasts were hunted down. Stability allowed trade to flourish more than ever before.

Harbors expanded to handle increased traffic. Stone breakwaters protected ships in storms. Watchtowers guarded coastal cities. Lighthouses guided sailors after sunset. Chains could be raised across harbor mouths to block enemy ships.

Rome took responsibility for maritime order, and the sea rewarded the empire with economic strength.

Ship Design Advances

As Rome gained experience, innovations continued. Shipbuilders refined hull shapes for better handling and speed. Sails improved to adapt to shifting winds. Steering systems evolved with stronger rudders.

Some ships were built to carry war elephants or siege towers. Others transported goods in bulk for trade. The navy also adopted techniques from conquered lands, especially Greek engineering.

Even rivers became part of strategy. Ships with flat bottoms could sail up shallow waterways, delivering troops where land armies struggled to reach. From coastline to coastline, Rome reshaped both the sea and itself.

Administration and the Business of the Navy

The empire built a detailed system to support naval operations. Registries listed every ship and crew. Supply officers managed food, tools, and medical supplies. Recruitment centers trained new sailors and rowers. Commanders coordinated fleets across regions, ensuring communication stayed efficient.

Inspectors reviewed ship conditions and crew readiness. Financial officers calculated costs and managed payment. The navy was not only built. It was managed with precision. This administration allowed Rome to field a navy that could outlast rivals during both peace and crisis.

A Psychological Weapon

The sight of a Roman fleet approaching a coastline inspired fear in enemies and confidence in allies. Ships carried more than weapons. They carried Rome's reputation for persistence and strength. When Rome arrived, it did so with preparation and intent.

Diplomacy often worked in Rome's favor because naval power stood behind every negotiation. When local leaders saw Roman banners rising from mastheads, few were eager to challenge them. Naval power became a quiet persuader across foreign shores.

Lessons Written in Wood and Iron

Rome's success at sea rested on a simple idea. When something must be learned, learn it quickly. When rivals showed strength, Rome copied and improved. When seas presented challenges, Rome built solutions.

The corvus eventually fell out of use because its weight made ships unstable in rough weather, but that did not lessen its role in early victories. Each innovation helped Rome take the next step toward long-term mastery. Over time, the navy grew more agile and professional. Rome never stopped adapting, and that constant evolution kept it ahead of threats.

The Lasting Tide of Power

In the shipyard, workers continued their tasks. Timber became hulls. Hulls became ships. Ships became legions upon the sea. Officers walked among the craftsmen with urgency and pride. They understood that Rome's future was rising plank by plank.

The transformation from land conquerors to naval rulers reshaped history. Rome no longer relied only on roads and infantry. Its reach extended across water, connecting continents and securing prosperity. Across calm waves and violent storms, Roman ships carried a civilization forward.

When the hammers finally fell silent and new warships slipped into the sea, they announced a new truth. The Mediterranean belonged to Rome. Control of its waters brought safety, wealth, and unity to an empire that would endure for centuries.

The shipwrights of Rome built more than vessels. They built the backbone of a world power.

Chapter 11
Beyond the Edge – Pacific Voyagers and Oceanic Navigation

The double-hulled canoe rose and fell with the long ocean swells, the water gleaming beneath a night sky crowded with stars. The voyagers sailed far beyond the sight of land, guided only by nature's signals. The navigator stood barefoot on the deck, feeling the movement of the waves through his legs. The canoe's sails caught the wind as the crew paddled gently to keep balance and speed. Above them, constellations traced invisible paths across the heavens. To outsiders, the sea seemed endless and dangerous. To these sailors, it was a familiar map, a place of direction, memory, and destiny.

The Pacific Ocean is the largest expanse of water on Earth. Across its vastness, ancient Polynesian navigators discovered and settled islands separated by thousands of miles. They did not rely on compasses or metal instruments. They did not possess written maps. Instead, they learned from the ocean itself. Every star, every bird, and every change in the swell carried information that guided them onward. Their voyages became one of humanity's greatest achievements in exploration.

Reading the Stars

For Polynesian navigators, the night sky was more than a backdrop. It was a living chart. They memorized rising and setting points of key stars along the horizon. Each star represented direction. When one disappeared, another guided the canoe forward.

Constellations taught time and distance. Some stars traveled higher overhead in certain seasons, signaling changes in weather or currents. The navigator tracked how fast stars moved across the sky to estimate speed. By knowing how long a star stayed in view, he understood how far the canoe had traveled.

Training took years. Navigators learned more than a hundred stars and their patterns. Knowledge passed through generations without writing. Words, stories, and songs preserved the secrets of the sky. The stars formed highways across the ocean.

Currents, Swells, and the Feel of the Sea

The Pacific contains distinct wave patterns created by storms far away. Swells move in different directions even when surface winds are calm.

Navigators read these movements with senses sharpened by experience.

Standing barefoot allowed them to feel subtle vibrations as waves struck the canoe. The direction of the main swell told navigators where they were in relation to known islands. Cross-swells, waves that traveled in a different direction, indicated a landmass causing the ocean to bounce back.

They watched changes in water color, reflections of clouds on the surface, and floating vegetation carried from shore. These signs confirmed that land was near. The ocean itself became a teacher.

Birds as Guides

Birds offered clues when land remained hidden. Navigators observed species that hunted fish near coasts or nested on islands. These birds flew out to sea in the morning and returned home each evening.

A small shape on the horizon could mean safety. Ground-nesting seabirds rarely slept on open water. If a flock appeared at dusk, the navigator followed them. Other birds indicated the direction of coastal shallows where fish gathered. Nature provided signals. Navigators interpreted them with skill and patience.

The Art of Finding Land

For long voyages, the navigator aimed toward a target island group rather than a single island. It was safer to reach a latitude and then sail east or west until land appeared.

Ocean spray, bird flight, cloud shapes, and floating debris offered hints. Clouds often formed differently over islands because land affected temperature and moisture patterns. Finding land required precision and confidence in knowledge passed through generations.

Canoes Built for Exploration

Polynesian voyaging relied on ship design as much as navigation. Double-hulled canoes offered stability and space for supplies. The hulls were lashed together with coconut fiber ropes that flexed naturally with the sea. Outrigger canoes used a stabilizer float to prevent capsizing. Both designs allowed fast and safe travel.

Craftsmen shaped hulls from logs and planks fitted tightly without metal nails. Sails made from woven mats caught steady trade winds. Canoes carried fresh water in gourds, coconuts, breadfruit, taro, and preserved fish. Some even supported gardens in small containers. Navigation and construction worked together to open the Pacific to exploration.

Oral Tradition: Libraries of Memory

Without writing systems, Polynesians relied on oral transmission. Navigational principles were taught through chants, stories, and star lore. Lessons were rehearsed until they lived in memory so deeply that forgetting was nearly impossible.

Chants described how stars traveled along the horizon. Myths explained bird behavior and ocean patterns. Songs encoded sailing directions to specific islands. Elders selected gifted students and trained them for years. Knowledge traveled with each voyage.

Reaching the Remote Pacific

Using these techniques, Polynesian voyagers settled islands across a vast region. They reached Rapa Nui in the east, Hawaiʻi in the north, and Aotearoa in the south. These journeys required crossing thousands of miles of ocean without land in sight for weeks.

Each island became a home where families planted crops and established communities. Yet deep connections to the sea remained. The entire region shared canoe designs, language roots, and cultural practices, revealing common origins. Despite long distances, they belonged to one oceanic family.

Mastery of Location Without Maps

Unlike many ancient sailors who stayed close to shore, Polynesians sailed boldly into open water. Their confidence came from training and a worldview that embraced exploration. The ocean was trusted. The stars were faithful. The navigator guided movement and belief.

Voyages were planned around seasons. Travelers waited for steady winds and favorable currents. They understood the risks but sailed anyway, certain the ocean offered opportunity. Courage and knowledge worked together.

Rediscovery of Ancient Knowledge

European conquest disrupted traditions. Canoe building declined. Some navigation skills faded. But in the 20th century, revival efforts began.

Modern navigators studied oral traditions and practiced reading stars and swells. In 1976, the Hawaiian canoe Hōkūle'a sailed from Hawai'i to Tahiti using only traditional navigation. The successful voyage proved the accuracy of ancient techniques. This renaissance restored pride and identity to Pacific cultures.

Community Across the Water

Oceanic navigation did more than move people. It created networks for food, art, and ceremony. Canoes made it possible to share traditions and maintain family ties across great distances.

Language carried stories. Fishing practices spread. Songs traveled. The Pacific became a shared home to cultures united by seafaring. Voyaging preserved connection.

A Sea of Memory

The canoe continued beneath the moonlit sky, slicing through water lit by starlight. The navigator's eyes moved from horizon to horizon, reading signs that only experience made clear. The ocean's pulse guided the vessel toward its destination.

To ancient Polynesians, the Pacific was not an obstacle. It was a pathway. They crossed the world's greatest ocean using only their senses and inherited knowledge. Their mastery showed that exploration does not require metal tools or written charts. It requires understanding and trust.

These voyagers stood among history's greatest explorers, proving that even the largest sea could be mapped by the human mind.

Chapter 12
The Compass of Empire – China's Maritime Pioneers

The slender needle quivered on its bronze plate as the ship rocked gently in harbor. Under the shade of fluttering red banners, a sailor tapped the side of the container, and the needle steadied, pointing in a fixed direction no matter how the ship moved. Officers nodded in approval. This tiny instrument promised confidence even when clouds hid the stars and land lay far beyond the horizon. On the deck nearby, workers hauled crates of porcelain, silk, and spices up the gangplank while musicians played to mark the beginning of a long voyage. China's sailors stood ready to explore distant waters, guided by forces hidden beneath the earth itself.

Across centuries, China became a leader in maritime innovation. While small fishing boats had once hugged the coastline, new technologies transformed Chinese seafaring into a bold global enterprise. The magnetic compass, advances in ship design, and large-scale naval expeditions revealed a civilization looking outward with ambition. The empire came to see the sea not as a threat, but as a bridge to the wider world.

Magnetic Curiosity Becomes a Tool for Navigation

Chinese understanding of magnetism began with lodestone, a naturally magnetic mineral. Scholars noticed that suspended lodestones rotated to align with a steady direction. During the Han Dynasty, inventors created a "south-pointing spoon," a smooth lodestone spoon that turned itself on a polished bronze plate. Although used mainly for fortune-telling and geography at first, this discovery hinted at new possibilities.

Craftsmen improved the concept over time. By the Song Dynasty, a needle rubbed with lodestone and floated on water or mounted on silk thread pointed reliably toward the magnetic north-south line. Sailors learned that this invisible force could guide them across open water when clouds blocked the stars or coastlines faded from view.

Writings from the 11th and 12th centuries describe the compass as a standard tool aboard ships. It did not replace star knowledge or wave reading, but it added security and precision. Navigators could travel farther without fear of losing direction. The compass changed what sailors believed was possible.

Ships Built to Travel Farther and Last Longer

As navigation improved, ship design advanced as well. Chinese shipbuilders experimented with large wooden vessels capable of surviving long voyages and fierce storms.

One breakthrough was the watertight bulkhead. Shipwrights divided the ship's interior into separate compartments with strong wooden walls. If one section leaked, water stayed contained rather than flooding the entire hull. This design saved countless ships.

Another major innovation was the sternpost rudder, mounted at the ship's stern. Sailors controlled it with a simple mechanism that allowed smoother turning and better handling in rough seas. This design appeared in China centuries before it was known in Europe.

Shipbuilders perfected balanced lug sails and used strong bamboo masts. Flexible, interlocking planks formed hulls able to withstand powerful waves. Some ships grew enormous, carrying hundreds of sailors and large volumes of cargo. Shipbuilding became a quiet revolution that turned distant seas into navigable highways.

Sea Routes of Trade and Diplomacy

Chinese sailors carried their goods and knowledge across Asia. Merchant ships from southern ports sailed to Southeast Asia, India, Persia, and the east coast of Africa. These voyages exchanged silk, porcelain, lacquerware, and metal goods for pearls, spices, ivory, and exotic animals.

Just as the Silk Road connected cultures across land, this maritime network linked peoples across the sea. Harbors in Guangzhou and Quanzhou became gateways to the wider world. Ships returned home with wealth and with stories of distant customs and technologies.

Sea trade strengthened China's economy and nurtured peaceful cultural exchange. It also suggested that control of ocean routes could spread influence far from the mainland.

Zheng He and the Treasure Fleets

During the Ming Dynasty, the emperor sought to display China's greatness with unmatched power. The admiral Zheng He, a skilled navigator and trusted court official, commanded massive fleets known as treasure fleets.

Accounts describe enormous treasure ships, some claimed to exceed 400 feet in length. Whether or not every measurement was accurate, the scale of these fleets was extraordinary. Hundreds of ships and tens of thousands of sailors traveled together like floating cities.

Between 1405 and 1433, Zheng He led seven global expeditions. His fleets visited Southeast Asia, India, the Middle East, and Africa. They returned with foreign envoys, rare animals such as giraffes, and valuable goods that amazed the court. Each voyage declared a simple message: China was a powerful, wealthy empire with vast reach.

Technology as Power

China's maritime success reflected a blend of technology and ambition. The compass, advanced shipbuilding, and refined navigation allowed Chinese sailors to journey with confidence when many other cultures still relied on coastal guidance.

While European sailors of the 1400s depended heavily on celestial navigation, Chinese mariners integrated magnetic direction into routine practice. Their ships carried innovations that other regions would adopt centuries later. Their naval organization supported voyages on a scale that astonished the world. China proved that control of the sea could protect trade, build alliances, and project influence.

Knowledge Shared Through Practice

Navigation knowledge did not remain limited to royal circles. Fishermen, merchants, and sailors improved their skills through experience. Sea charts known as "rutters" recorded currents, hazards, and landmarks. Sailors passed these guides along with stories of storms, shoals, and pirate activity.

Compass designs became more precise. Markings allowed sailors to measure direction accurately. Over time, the tool spread along Indian Ocean trade routes and later influenced navigation across the globe. Innovation grew through daily practice.

Comparisons Across the World

Different regions excelled in unique maritime traditions. Polynesian navigators crossed vast oceans using stars and swells rather than instruments. Indian Ocean sailors mastered the monsoons. Mediterranean powers dominated naval warfare and harbor construction.

Each tradition shaped history in its own way. Yet during Zheng He's time, Chinese fleets represented the most advanced combination of ship size, navigation, and state coordination. Their achievements stood at the crossroads of science and diplomacy. Had the voyages continued, global exploration might have unfolded differently.

A Turn Inward

After Zheng He's final voyage, political priorities shifted. Inland concerns and economic pressures led officials to scale back maritime ambition. The great fleets were docked or dismantled. Overseas exploration slowed.

Yet coastal communities continued to sail. Private merchants kept small-scale trade alive. Navigation skills survived in practice rather than in grand expeditions. Even when priorities shifted, curiosity endured.

The Compass Guides the Future

The magnetic compass traveled far beyond China. European sailors adopted it and refined it, relying on it heavily during the Age of Exploration. Ship designs such as the sternpost rudder spread westward and influenced fleets that later crossed the Atlantic.

Chinese innovation helped ignite global seafaring. A simple stone and metal experiment evolved into a tool that reshaped world navigation. Knowledge from Chinese shipyards changed every ocean-crossing empire that followed.

Endless Possibility on the Water

The ship in harbor prepared to leave under the guidance of the small, unwavering needle. Sailors pushed away from the pier as banners snapped in the wind and the rudder turned the vessel toward open water. They could not predict what lay across the sea, but they carried tools and training that brought confidence.

The ocean remained wide. Challenges lingered beyond the horizon. Yet experience and innovation made those challenges manageable. China's maritime pioneers embraced both tradition and invention.

For them, the sea was not a barrier. It was the beginning of a journey. The compass proved that no distance was too great for those determined to explore.

Their sails revealed that progress could follow paths guided not only by sight, but by understanding the natural forces connecting every shore.

Chapter 13
Bronze, Bits, and Bridles – Animal Power and Speed

The horse tossed its head as the craftsman lifted the bridle. Two handlers steadied the animal, leaning their shoulders into the ribcage while a third soothed it with a calm voice. Morning light washed over a courtyard of tools and leather. The metal bit flashed once before the craftsman guided it past the fleecing lips and onto the tongue. The horse's jaw clenched and relaxed. Steam rose from its nostrils. When the cheekpieces settled and the throatlatch was buckled, the animal's restless steps changed. The handlers stepped back. A tug on the reins sent a clear message through bronze and fiber. The horse's ears flicked forward. Control and speed, once at odds, were now partners.

Around them, saddles rested on wooden racks. Pads were stuffed with wool and horsehair. Stirrups hung like small iron doors, polished smooth by years of use. In a corner, ropes, nosebands, and rawhide straps lay coiled and ready. Every piece seemed simple. Together, they formed a system that would carry messages, soldiers, and ideas across continents.

Harnessing the First Speed

The story of animal power began long before metal. People learned to live with animals and shape their work, turning strength into motion. Donkeys and onagers hauled loads across river valleys. Oxen pushed plows through heavy soil and dragged sledges over rough ground. Horses, first kept for meat and milk on the grasslands, slowly became partners in movement. The steppe taught humans that speed was not only a measure of running, but of trust, training, and the right touch on a rope.

Early riders did not sit in carved saddles or hold refined bridles. They guided animals with nose rings, halters, and simple mouthpieces made from rope, bone, or wood. A skilled rider learned the feel of a back under the hips, the spring of muscle, and the warning signs of fear. The first steps were slow, but once people could stay on and steer, the world grew smaller.

Metal in the Mouth

Bronze bits arrived as craftsmen turned ore into tools that could send finer signals. A metal bit did not tire like a rope. It delivered pressure evenly, returned to shape, and transmitted a rider's intent with clarity. Side rings stabilized the mouthpiece. Jointed sections allowed the bit to fold and communicate in stages, from a whisper to a command. These details gave riders dependable control at speed, and speed demanded control most of all.

Iron followed bronze, stronger and easier to shape in thin forms. Cheekpieces became more elaborate. Some bits introduced leverage that multiplied the pull on the reins. Good horsemen used that power with care, because harsh hands dulled a mount and ruined trust. The best riders needed less force, not more. Metal expanded the language between human and animal, but it did not replace patience or practice. It made communication consistent even when sweat and fear complicated the message.

Saddles and the Rider's Seat

A saddle seems like a cushion at a glance, but in truth it is a device that manages force. Early pads spread the rider's weight, making long rides possible without scraping skin or crushing muscle. As riding grew more demanding, trees—frames of wood covered in leather—lifted the rider off the spine and placed the legs where they could act without hurting the horse. A well-fitted tree created space along the back where tendons could stretch and lungs could fill.

Straps anchored the saddle against sliding or twisting. A breastcollar kept it from drifting toward the rump on climbs. A crupper prevented it from creeping forward on descents. With these adjustments, a mount could trot all morning and still gallop when called upon. Saddles protected bodies on both sides of the relationship, allowing endurance to become strategy.

The Invention of the Stirrip

The stirrup did something simple that changed almost everything. It gave the rider a platform. With the ball of the foot supported, the legs could bear shocks, the hips could rise and sit again, and the hands could guide rather than cling. A rider could stand to shoot an arrow or lean forward to absorb a charge. Horses leaped, pivoted, and accelerated under a pilot who no longer fought to stay mounted.

Stirrups did not appear everywhere at once. Riders in different regions solved balance with knees, tall saddle bows, or long practice. But when metal and leather loops spread across Eurasia, they tied the rider to the animal with new stability. Cavalry could hit harder and recover faster. Couriers could cross rough terrain with fewer falls. By holding the foot in place, the stirrup freed the rest of the body to work.

Messages That Outran Armies

Once riding became reliable, messages moved faster than marching feet. Mounted couriers stitched empires together with time. A sealed tablet left a capital at dawn, reached a border town by sunset, changed horses at a relay station, and sped onward through the night. The road mattered, but the horse mattered more. Fresh mounts, water, and grain placed at intervals formed an invisible machine that turned distance into schedule.

Governments learned to measure loyalty by speed. Reports arrived while events still mattered. Governors received orders in weeks instead of months. Traders traveled the same paths with receipts and contracts tucked into belts. Intelligence returned with couriers who stored landscapes in memory. The mounted post made information a living thing, always in motion.

Cavalry Remakes the Battlefield

On the battlefield, the horse transformed war. Infantry could hold ground. Cavalry could choose where the fight happened. Light riders swarmed the edges, shooting and fading like shifting wind. Heavier riders, protected by layered armor and mounted on strong-boned horses, gathered into wedges that drove through lines. Cataphracts—horse and rider armored as one—pressed forward like moving walls. Horse archers sketched circles around slower forces, releasing arrows at a canter and breaking away before spear points could reach.

These tactics required gear that matched the task. Bits steady enough to keep lines true at speed, saddles firm enough to hold riders on impact, and stirrups that steadied bows or lowered spears at the final heartbeat before contact. On plains and in valleys, cavalry often decided whether a campaign ended that day or endured for years. Speed magnified courage and fear at the same time.

Across Sand and Salt

In deserts and dry steppes, the camel took the lead. A horse drank often and needed pasture. A camel, shaped by heat and scarcity, turned arid land into a highway. Its step was long and swaying, its feet wide and soft to spread weight on sand. Thick lips plucked thorny shrubs. Nostrils closed against dust. Water carried in skins and fat stored in the hump stretched the range between wells.

Saddles for camels solved different problems than for horses. A high wooden frame could straddle the hump, balancing loads while keeping the spine free. Riding saddles sat behind or over the hump, giving the rider a high view across dunes. With this equipment, caravans moved thousands of pounds of incense, textiles, and grain along routes where wagons failed. Guides knew the taste of each well and the shape of each dune ridge. Their maps lived in memory.

Culture in the Stable

Animal husbandry reflected what people valued and what the land allowed. Steppe herders moved with seasons, reading grass like a script. Foals grew sure-footed in open country. Mares' milk and meat supported families when crops could not. In wetter regions, farmers bred compact horses for plowing and hauling. In cities, stables sat behind courtyards with thick walls against winter winds. Grooms rose before dawn to brush coats, check hooves, and carry buckets. Care became craft, and craft became pride.

Training practices varied. Some broke horses quickly with force, producing obedience but dullness. Others taught the animal to accept bit and saddle with less fear. Songs and steady routines calmed young stock. In arid lands, camel calves learned to rise under load and take the lead rope without injury. Every region mixed patience with pressure, and the balance reflected culture as much as need.

Nomads and Cities Meet

As mounted speed expanded, borders blurred. Nomadic riders arrived at town gates with wool, hides, and livestock to trade for grain, metal, and tools. Markets filled with bargaining voices. News traveled on saddles and returned with prices. Yet friction grew. Pasture needed space. Fields needed fences. Kings tried to tax movement. Clans resisted orders shaped by settled calendars.

Sometimes interaction meant raiding or defense towers. Sometimes it meant intermarriage and shared rituals. Mounted patrols from settled states entered grasslands to show presence. Nomad bands escorted caravans for pay. The frontier became a broad zone where hoofprints overlapped. Over time, farmers learned to ride better and riders learned to keep records. Slowly, speed reshaped identity.

The Postal Mind

With horses, states learned to think in distances measured by hours. Relay posts demanded schedules. Stable masters tallied feed. Blacksmiths measured iron for shoes. Roads adapted to animal needs. Steep climbs became switchbacks. River crossings shifted to shallow bends. Inns appeared where riders expected to stop. At dusk, tired riders sought stew and sleep while tired horses sought grain and water.

The postal system standardized language. Place names appeared in official orders the same way each time. Landmarks became common knowledge. Calendars mattered more when transportation kept time. Mounted networks gave rulers the illusion of presence. A decree sent at noon could arrive far away by the next sunrise. That reach encouraged ambition.

The Price of Motion

Speed carried costs. Pasture near towns thinned. Grazing routes crossed fields and sparked disputes. Iron, leather, and fiber wore out faster when miles multiplied. A lame horse could end a mission. A broken girth strap could injure a rider. Maintaining many animals required water management, forage planning, and skilled labor. Winter hay had to be cut and stacked. Summer work began before sunrise and ended by lantern light.

There were human costs as well. Cavalry raids left empty homesteads. Couriers drowned at river crossings or vanished in storms. Camels sank in salt flats when crusts collapsed. Tools increased human reach but did not erase risk. People accepted the bargain because the returns—victory, trade, connection—were great.

Technologies That Travel

Bits, saddles, and stirrups crossed borders quickly. A traveler admired a neighbor's bridle and copied the cheekpiece shape. A captured saddle returned home as a pattern for new designs. Traders carried small iron loops and explained how to adjust their length. Over time, gear blended from region to region. A noseband from the east met a saddle tree from the west and produced something new.

Diffusion moved in many directions. Desert saddles shaped steppe designs. Steppe bow cases appeared on city saddles. Camels carried horse gear along with spices. Markets mixed solutions as easily as goods.

Animals Beyond the Horse

Not every landscape favored horses. In northern forests, reindeer pulled sledges across snow. In high plateaus, sure-footed mules carried packs along cliff paths too narrow for wagons. In marshlands, water buffalo hauled plows through mud that swallowed lighter animals. Each region built its own balance of speed and strength. The lesson stayed constant: tools that fit the land and the animal multiplied power.

Even within horse cultures, breeds diverged to match work. Some favored long legs and deep lungs for distance. Others grew heavier for carrying armor. Coat colors carried ritual meaning, but the real measure lay in how a horse took the bit, held the line, and returned home sound.

Speed as Symbol

The meaning of riding expanded with its use. A mounted messenger represented a state that could reach the farthest corners. A cavalry officer in parade armor reminded people that peace rested on controlled force. In town squares, races displayed breeding and training. Children ran behind the competitors until their chests hurt, pretending to be the riders they watched.

Poems praised horses that did not tire and camels that walked like ships across sand. Tombs depicted riders mid-gallop. Saddles were gifts that sealed alliances. Metal bits hung near doorways as signs that the family inside knew how to move and how to return.

Communication Remade

When words could cross a province before rivers froze, negotiation changed. A grain shortage in a distant district no longer became a prophecy of revolt. A message went out. Wagons rolled. Pack strings formed. Need eased before anger ripened. When enemies gathered, scouts on lean horses counted forces and delivered the report before the threat arrived. Leaders chose with better timing, and sometimes that timing saved lives.

Merchants benefited. Prices shifted smoothly when information moved quickly. Traders avoided flooded valleys because riders reported the danger days earlier. Caravans met escorts at wells because letters promised they would. Exchange gained a rhythm that followed hooves.

What Animals Taught People

The long partnership changed its participants. Riders learned to think with legs as well as hands. They measured distance in hours, listened to weather with skin, and understood another being's limits. Handlers learned that care at dawn—water, forage, a check for soreness—mattered as much as courage at noon.

This discipline spread into daily life. Schedules mattered. Preparation mattered. One small failure could undo a season's work. Animal power hardened communities and also taught tenderness for the lives that made speed possible.

Meeting the Horizon

The horse in the courtyard stood quiet now, the bit warmed by breath. The handlers let the reins fall slack for a moment. The craftsman checked the fit one last time, feeling for pinches along the lip. Satisfied, he nodded to the waiting rider. The saddle settled with a soft thud. A foot found the stirrup. Weight rose. The horse walked, then trotted, then eased into a careful canter toward the open gate.

Beyond that gate stretched fields, roads, deserts, and steppe. Somewhere, a courier traded mounts at a relay post and rode into the night. Somewhere else, a unit tightened girths before a charge. In a distant market, a camel knelt as bales of cloth slid to the ground. In a nearby pasture, foals raced in loose circles, discovering the first joy of their own speed.

Bits, stirrups, and saddles did not create motion. They shaped it, turning strength into direction and distance into time that could be planned. With metal and leather, with hay and water, people made a partnership that changed how far they could reach and how fast they could respond. Mobility linked ways of life, forcing nomad and farmer, sailor and soldier, to meet until exchange and conflict wrote a shared history on the same roads. The world learned to move. And once it learned, it did not forget.

Chapter 14
When the Wind Fights the Sea – Sails, Rigging, and Naval Reach

Wind smashed into the canvas like a living creature, snapping lines taut and forcing sailors to scramble across the deck. The ship heeled sharply as spray leapt the rail. Ropes hissed through wooden blocks while shouts carried over the roar of the sea. One quick haul, one well-timed adjustment, and the sails that strained to break free instead filled perfectly, transforming wild force into forward motion. The ship's bow sliced through the waves, climbing each crest with determination. Every sailor understood in that moment that mastery of wind and rigging was not only a matter of survival. It was the power to move farther, faster, and with purpose.

Wind could be a friend or a destroyer. Sails determined which role it played.

The Craft of Capturing the Sky

Early sailors relied on simple square sails set on single masts. These sails worked well when wind blew from behind, pushing ships forward like leaves on a stream. The Mediterranean's reliable seasonal winds supported these early voyages. Rafts and small wooden boats crossed short distances, carrying traders from coast to coast. Control was limited, but the idea was proven: harnessing wind created a highway across water.

As seafarers ventured farther, they demanded more control. The square sail faltered whenever wind shifted or blew against the bow. If ships could not adjust, they drifted backward or sideways. Sailors experimented with how the sail hung, trimming corners, or angling the mast. Each discovery added a new note to the language of the sea.

These improvements turned sails into precision tools. A sail was not only a sheet catching wind. It became a method of steering the future.

Rigging and Lines: The Hidden Machinery

Sails alone could not command a ship. Lines, twisted from plant fibers, became the nervous system connecting canvas to crew. Blocks and pulleys multiplied muscle power. Ratlines made ladders to the sky. The more complicated the rigging grew, the greater the need for skill and coordination.

On deck, sailors stood ready to haul, slacken, or belay at a moment's notice. Above them, riggers balanced on swaying masts. Every rope had a name: halyards to raise sails, sheets to adjust angle, braces to swing the yards. Mistakes could send a man flying or tear a sail to ribbons. Storms did not forgive confusion. Clear commands and practiced teamwork kept ship and cargo safe.

Rigging did more than control motion. It forged community under pressure.

Learning to Climb the Wind

A major leap in sail evolution came with tacking and beating against the wind. By angling sails sharply and letting air slide along the canvas, sailors turned wind that resisted them into wind that helped them. Sharp hulls cut deep into the water, and the keel kept ships from sliding sideways.

This skill opened new routes. The limits of the sea shifted with human understanding. Navigators plotted ambitious paths along coasts and across open waters. Return journeys no longer depended on waiting for wind to change. Sailors chose their direction instead of hoping for it. Knowledge turned necessity into freedom.

Multiple Masts, Longer Voyages

Adding more masts gave sailors flexibility and strength. Fore-and-aft sails worked best near the wind; square sails provided power with wind behind. Combining them allowed ships to respond quickly to shifting conditions.

Larger hulls required more than one mast to drive them forward. Extra canvas meant redundancy. If one sail tore or a mast snapped, the others kept the vessel moving until repairs could be made or shelter reached. Longer voyages became realistic, and ships carried heavier cargo without losing speed. The horizon became not an edge, but an invitation.

Exchanging Ideas from Shore to Shore

Sailing knowledge traveled along the same routes it enabled. Maritime societies studied one another. Egyptians used simple square sails on the Nile. Phoenicians perfected multi-decked merchant ships. Greek and Roman sailors improved rigging and hull design. In the Indian Ocean, Arab navigators sailed dhows with triangular lateen sails that sliced neatly across the wind.

When these cultures met through trade or war, they copied what worked. Lateen sails entered Mediterranean rigs, producing hybrids that outperformed either tradition alone. European ships later adopted stern-hung rudders and refined rigging from Asian models. Every improvement became a shared achievement. The sea was not a barrier to innovation. It was the quickest path for ideas to move.

Scientific Thought Reshapes Seafaring

As centuries passed, sailors and scholars studied why ships moved as they did. Wind patterns revealed the shape of global climate. Buoyancy explained how hulls floated and how to load them safely. Shipwrights sharpened bows for cutting, widened sterns for stability, and reinforced frames with strong timbers.

Carpenters built ships like puzzles. Curved planks overlapped in clinker style or met edge-to-edge in carvel construction. Caulking sealed joints. Resin and pitch protected wood. Rope makers twisted tougher fibers. Block makers shaped pulleys smooth enough to glide under strain. Sailing was physical, but its future grew from understanding.

Markets Expand with the Wind

Better sails meant better trade. Larger ships carried more pottery, olive oil, wine, and grain. Ports expanded with warehouses and customs houses. Cities grew rich by feeding distant lands and importing luxury goods. Merchants no longer served only neighbors but customers across the sea.

Transport costs fell as capacity rose. Armies marched with lighter loads because ships carried supplies faster. Farmers planted new crops knowing ships would deliver harvests to faraway appetites. Economic power shifted from land routes to sea lanes. Commerce followed the wind.

Naval Warfare Takes a New Shape

Speed and maneuverability transformed warfare at sea. Warships armed with rams required precise control, turning sharply to strike enemy hulls. Later, ships mounted towers and artillery, turning decks into platforms for long-range attacks. Fleets drilled in formations where one command altered the path of dozens of vessels.

Sails gave commanders reach. A navy could protect island chains, enforce taxes on sea trade, or blockade rival harbors. Empires gained confidence not only in walls and soldiers, but in their sails. Power rested on wood, canvas, and wind.

Sailors Who Knew the World

Life at sea demanded mastery of everything the ocean offered. Sailors read weather in clouds and air. They knew which currents would speed or slow their progress. They understood the moods of waves and the safety of harbors.

Though tools sharpened and ships grew taller, the sailor's life remained hard. Storms shattered masts. Sails tore loose. Rigging whipped across decks. But for those who survived, the sea offered pride. Every safe return proved skill and readiness to go again. To sail was to become tougher and wiser.

Ports as Crossroads of Culture

Harbors became places where worlds converged. Sailors swapped words, melodies, and meals. Navigators compared notes on reefs and shoals. Shipbuilders studied foreign hulls. Merchants sampled spices and wondered what fortune lay beyond the horizon.

Some ports became symbols of connection—Athens with its Long Walls, Carthage with its circular harbor, Alexandria with its lighthouse. Each announced that ships came from and went to everywhere. Empires rose by claiming these crossroads.

Sails Change Identity

Governments measured strength by fleet size. Maps centered on coastlines rather than capitals. Craft guilds specialized: ropewalks stretched hemp into miles of cordage, while sail lofts cut and stitched canvas into powerful shapes. Art celebrated the sea's growing influence. Coins bore ships under full sail. Tomb reliefs showed naval battles. Poets wrote of voyages that brought glory or ruin.

The ship at sea became a symbol of national ambition.

Holding the Wind in Cloth

The merchant vessel from earlier steadied as its crew finished trimming the sails. The force that had threatened to capsize it now drove it forward. Canvas bulged under tension. Masts hummed. Water parted cleanly along the hull. The sailors rested a moment, breathing salt air, then returned to work.

Above them, the sky remained vast and unpredictable. Storms brewed without warning. Winds shifted from gift to danger in a heartbeat. But for those who understood atmosphere and ocean, fear coexisted with purpose.

The evolution of sails and rigging turned wind from obstacle into opportunity. Knowledge spread with every voyage. Markets expanded. Armies marched with supplies carried by sea. Civilizations stitched themselves together with thread spun from plants and strength carved from trees.

With each improvement, sailors stepped farther from shore, confident that wind—though wild—could be shaped to serve. As long as canvas caught the air and hulls cut the waves, human reach stretched across oceans, proving that the sea's power could be guided by skillful hands and well-drawn lines.

Chapter 15
Bridges Across Kingdoms – Spanning Rivers and Ravines

Ropes strained under the weight of stone as builders pulled and braced along the edge of a deep ravine. The ground fell away into jagged cliffs where a narrow river foamed below. Workers shouted instructions while a master builder paced the near bank, measuring with his eyes the distance that must be crossed. On the far side, another team hammered poles into place to anchor supports. Dust rose with each strike of the mallet. The sun beat down on crews who risked falls and failure for a promise: when this structure stood, people could walk from one land to another in a single, confident stride. Villages would connect. Traders could pass without waiting for a boat or daring treacherous waters. A ravine that once marked separation would become a seam joining two worlds.

Every ancient civilization faced the same problem. Rivers blocked movement. Gorges cut through rock.

Seasonal floods erased paths. To expand, trade, and defend, people needed crossings they could trust. Bridges turned natural barriers into pathways of power.

The First Steps Across

Long before stone arches or suspension designs, people built simple crossings. Fallen trees served as the earliest bridges. Logs rolled into place turned muddy banks into passable routes. Over time, planks improved balance and safety. Wooden beams laid across narrow streams allowed herders, travelers, and merchants to continue without delay.

Communities experimented with supports—posts driven into riverbeds to hold wider decks. Rope handrails kept feet steady. Even these small structures required planning: weight had to be balanced, water flow had to be considered, and wood needed protection from decay. Each improvement pushed bridge building from chance to calculation.

Symmetry in Stone

As techniques advanced, stone became the preferred material. It resisted rot, fire, and storms. Stacked dry or with mortar, stone carried immense weight. The breakthrough came with the arch.

An arch redirects downward force outward along its curve. Each stone locks its neighbors in place. The keystone prevents collapse, and side supports absorb pressure. This principle allowed spans longer and stronger than anything achieved before.

The arch did not spread evenly across the ancient world, but where it took hold, it reshaped engineering and empire.

Rome's Legacy of Connection

No ancient culture used stone arches more systematically than Rome. Roman engineers treated bridges as parts of a greater mobility network. Roads radiated outward from the capital, and bridges carried them across rivers and plains. The Tiber, Po, and Rhine—once obstacles—became elements of Roman reach.

The Pons Fabricius in Rome, built in 62 BCE, remains in use today. Its two arches rest on a solid stone pier and have served pedestrians for more than two millennia. The Alcántara Bridge in Hispania, spanning the Tagus River, bears an inscription dedicating it to Emperor Trajan. Its tall arches and broad roadway show the scale of Roman ambition.

Bridges supported soldiers, market supply chains, tax collectors, and imperial authority. For Rome, a bridge was not only construction. It was command.

Water's Force and the Challenge of Durability

Engineers learned that rivers are anything but passive. Floods could sweep away whole crossings. Ice could crush supports. To resist nature, builders shaped piers to cut water rather than block it. Diamond-shaped or rounded upstream faces split currents and reduced damage.

Foundations demanded precision. Workers drove piles deep into riverbeds to keep piers stable. In some regions, engineers built cofferdams—temporary enclosures that allowed them to pump water out and work on dry ground. These methods required teamwork in dangerous conditions where one misstep could mean drowning.

Bridging meant wrestling with unpredictable water as much as shaping wood or stone.

The Heights of the Andes: Inca Ingenuity

Elsewhere, geography led to different solutions. The Andes offered cliffs, fast rivers, and earthquakes. Heavy arches were impractical, so Andean builders turned to flexible materials.

The Inca braided ichu grass into thick ropes that became suspension cables. Weavers formed ropes as long as a city street, then anchored them to carved stone abutments. A planked walkway hung beneath, swaying above roaring water. The bridge flexed with use but rarely failed abruptly. Communities replaced frayed cables through the labor system known as mit'a.

These bridges connected the Inca road network and allowed chasqui runners and armies to cross chasms that would have stopped wheeled carts.

Crossing required trust in neighbors, craft, and a shared understanding that even the deepest gap could be conquered.

Paths for Armies and Ideas

Bridges reshaped how rulers imagined borders. Without reliable crossings, rivers acted as defenses. With them, aggression could flow both ways. Military leaders learned that bridges were strategic targets. Julius Caesar famously built a timber bridge across the Rhine in 55 BCE to prove that Roman power could reach Germanic lands. It took ten days—a calculated display of might.

Bridges also accelerated communication. Merchants no longer waited for ferries. Caravans crossed without unloading. Messengers delivered news in hours rather than days. Speed became power.

Trade expanded when river delays vanished. Goods traveled farther, markets grew, and cultural exchange quickened. Silk moved from East to West. Spices reached cities hungry for new flavors. Wool and metals found distant buyers. Bridges enabled the routes that connected early global commerce.

Symbols of Unity and Triumph

Bridges did more than carry travelers. They made statements. Their construction showed that a ruler commanded labor, resources, and expertise. Inscriptions praised patrons. Sculptures and reliefs decorated piers and parapets.

Crossing a great bridge could inspire awe. People looked up at sweeping spans and felt pride that human hands could tame distance and water. Ceremonies accompanied openings. Festivals marked the achievement. A bridge declared: "We bring you together. We make your world larger."

Different Materials, Shared Purpose

Across continents, bridge builders solved similar problems in different ways. In Asia, Chinese engineers built carved stone-beam bridges and sweeping arches like the Zhaozhou Bridge, whose segmental arch reduced weight while maintaining strength. In India, teak bridges spanned streams. In Japan, sacred bridges marked entrances to temple grounds.

In the Middle East, wooden spans crossed wadis that ran dry most of the year but flooded suddenly. In Europe, early medieval builders preserved Roman methods where they could and improvised where they could not.

Designs differed, but the purpose remained constant: movement strengthens life.

Wood and Rope Where Stone Would Fail

Not every region had abundant stone or labor. Forest-rich lands favored timber trusses that angled beams to distribute force. These bridges required maintenance and were vulnerable to rot, pests, and fire. Yet they were quick to build, easy to repair, and adaptable.

Rope bridges appeared in mountainous regions beyond the Andes. In parts of Asia, bamboo served as strong cable material. Lightweight structures could hang over deep ravines that heavier stone spans could never reach. Travelers learned to move in rhythm with the bridge's sway, trusting motion instead of resisting it. Adaptation made success possible.

Crossings Define Connections

Bridges influenced where towns and cities grew. A settlement near a reliable ford might stay small, but a settlement near a major bridge often rose to prominence. Inns, warehouses, and markets clustered around these strategic points. Tolls filled treasuries. New roads aligned with the best crossings, reinforcing their importance.

Some bridges served as neutral trading grounds between rival lands. Others became flashpoints when passage was blocked or taxed heavily. Control of a bridge often meant control of wealth.

Conflict and Collapse

When war erupted, bridges became targets. Armies removed planks, burned wooden spans, or demolished piers to slow pursuit. Floods destroyed crossings in minutes. Earthquakes shattered arches. Repairs consumed resources long after battles ended.

Engineers responded with stronger piers, better understanding of river behavior, and improved reinforcement. Yet every bridge required constant care. Neglect could sever connections as surely as war. Work that holds the world together requires steady hands.

Engineering as Knowledge

The math behind bridges grew more precise with each generation. Builders refined understanding of compression and tension. They learned that deeper arches carry greater loads, wider piers grant stability, and longer spans demand careful material choices.

Geometry supported success. Accurate measurements prevented twisting decks and unbalanced pressure. Water levels aligned piers. Plumb lines ensured vertical accuracy. Templates standardized arch stones. Each project added to a growing foundation of structural knowledge. Bridging advanced one span at a time.

Rituals of Crossing

In many cultures, crossing a bridge held symbolic meaning. Travelers passed from one community to another, leaving behind one set of customs and entering another. Brides crossed bridges during wedding processions. Kings rode across new structures to claim authority. Religious leaders blessed crossings to protect all who walked over them.

Bridges became metaphors for unity and transition. To burn a bridge means to sever ties. To build one means to restore connection. Physical crossings shaped mental ones.

A Future Built on Crossings

The workers hauling ropes and stones at the ravine's edge labored despite heat and danger. They knew that once the bridge stood, the lives of everyone nearby would change. Travel would quicken. Markets would grow. Families split by terrain would reunite. Opportunities would emerge where none existed before.

Long after builders departed, the bridge would continue its quiet work. Children would run across it without imagining the earlier risk. Merchants would count profits enabled by its span. Soldiers would march over it without noticing the hands that shaped each stone. Bridges made movement normal. They made connection expected.

The World Spans Its Gaps

As centuries passed, designs evolved—wider decks, higher arches, stronger supports—but the purpose remained unchanged: to join where nature divides. Ancient bridges endure today not only as feats of engineering but as reminders that humans refuse to accept isolation as fate.

A river or ravine may appear final, but a bridge insists it is temporary.

With rope, timber, stone, and skill, humanity learned to overcome distance. Bridges became physical proof that imagination and cooperation could redraw maps. They turned barriers into opportunities and reshaped the ancient world into a lattice of paths connecting one kingdom to the next. The moment a bridge links two lands, both lands change forever.

Chapter 16
Ports of Exchange – The Ancient World's Global Gateways

The harbor teemed with motion and noise. Ships with tall masts and striped sails nudged against wooden wharves as sailors tossed thick ropes to workers waiting on shore. Cargo handlers shouted instructions as crates, jars, and woven bundles swung through the air on creaking cranes. The tang of salt mixed with the smell of spices spilling from baskets. Merchants argued in a dozen languages, counting coins, praising goods, and striking deals with quick hand gestures. Priests moved through the crowd offering blessings for safe passage, while scholars recorded cargo lists in neat script. Behind the piers, warehouses rose like guardians of wealth. A lighthouse towered over the scene, its flame visible to ships still days away. In this place, the sea delivered more than goods. It delivered new ideas, new foods, new beliefs, and new ways of living.

Ports were more than waterfronts. They were engines of change where distant worlds met and forged connections that reshaped civilizations.

Gateways to Empires

As maritime technology advanced, certain harbors emerged as critical hubs. Geography determined their rise. A port aligned with reliable winds or major sea routes became a magnet for trade. Once ships gathered there, improvements followed: deeper docks, stronger piers, guarded storage yards, and secure anchorages. Cities formed behind these waterfronts, growing wealthy as goods and travelers passed through.

Each port developed a character shaped by its environment. Some specialized in luxury goods. Others fed inland settlements with bulk cargo. All served as meeting points where cultures mingled, creating societies unmatched in diversity.

Alexandria: Where Knowledge and Commerce Met

Alexandria in Egypt stood as one of the brightest centers of the ancient world. Founded by Alexander the Great and expanded under the Ptolemies, it commanded the entrance to the Nile and the gateway to the Mediterranean. The Pharos lighthouse, one of the Seven Wonders, signaled safety and ambition. Its bright flame guided ships through dangerous shoals, saving lives and securing profit.

Alexandria's harbors stored grain that Egypt exported to the wider Mediterranean. Markets overflowed with papyrus, linen, and jewelry. Scholars from Greece, Persia, and beyond gathered at the Library and the Museum, where the world's knowledge was collected and studied. Lecture halls rang with debates in philosophy, mathematics, and astronomy. Scientific instruments arrived in crates alongside spices and fine glassware.

Here, ideas crossed borders as easily as ships crossed water.

The port was where books traveled, where languages blended, and where invention thrived.

Carthage: A Crossroads of Wealth and Rivalry

Across the Mediterranean, Carthage rose as a powerhouse of trade. Its protected harbor, designed with concentric basins, separated military and commercial fleets with remarkable organization. Merchants sailed along the African coast, past Iberia, and into the Atlantic, forming a vast network.

Carthage traded metals, salt, and agricultural goods. Purple dye from murex shells—worth more than gold—made the city famous among elites. Carthaginian sailors mastered daring long-distance voyages along the West African coast in search of new markets and resources.

Prosperity, however, drew conflict. Rome eyed Carthage's wealth with growing envy. Their rivalry erupted into the Punic Wars, where naval battles and sieges determined who would control Mediterranean trade. Carthage's ports became strategic prizes. Its rise and fall proved that control of harbors could determine the fate of empires.

Ostia: Rome's Lifeline to the Sea

Rome ruled the land, but without a port, it would have starved. Ostia filled that role. Located at the mouth of the Tiber, it served as the arrival point for grain shipments from Egypt and North Africa. Warehouses stored wheat, olive oil, wine, and fish sauce that sustained millions.

Docks were organized with impressive efficiency. Workers unloaded amphorae and stamped them with symbols showing origin and quality. Inns, taverns, and baths lined the waterfront. Bureaucrats checked manifests, and tax collectors claimed their share. A lighthouse guided ships to safe harbor, acknowledging the sea as both partner and threat.

Ostia was Rome's pantry, customs office, and gateway to distant peoples. It lacked Alexandria's grandeur and Carthage's ambition, yet without it, Rome's greatness might not have endured.

Guangzhou: Where the Silk Roads Met the Sea

Along China's southern coast, Guangzhou—historically Canton—became a major ocean port. It connected inland Silk Road routes with maritime networks stretching through Southeast Asia, India, Arabia, and East Africa. Junks, dhows, and other vessels crowded the harbor. Cargoes of silk, porcelain, lacquerware, salt, and metal goods flowed toward distant buyers.

Foreign traders lived in special districts, bringing unfamiliar languages and customs. Mosques, temples, and trading halls stood side by side. Officials monitored transactions, collected taxes, and enforced laws. The port became a laboratory of cultural exchange, where navigation knowledge, medicine, and mathematics traveled between worlds.

Guangzhou's success showed how sea trade enriched both sides of exchange, not only in wealth but in wisdom.

Tools That Shaped the Harbor

Ports needed infrastructure to function well. Docks provided platforms for loading and unloading. Wood and stone supported heavy cargo and allowed access at shifting tides. Cranes, powered by muscle or simple machines, lifted goods too heavy for human strength alone. Lighthouses warned sailors away from hazards and guided them toward safety.

Customs offices tracked goods and collected duties. Market squares sat near water so trade could begin immediately. Inns sheltered traveling merchants who carried news from faraway lands.

Each improvement reduced transport costs and brought distant societies closer.

Ports as Melting Pots of Culture

The most remarkable feature of ancient ports was not their buildings—it was their people. Merchants, sailors, diplomats, refugees, and scholars gathered in crowds that mixed religion, fashion, and music.

A traveler arriving in Alexandria might hear Greek debate alongside Egyptian prayer. In Ostia, Latin mingled with Iberian, Gallic, and Punic tongues. Guangzhou buzzed with sailors from Java, India, and Arabia.

These encounters shaped new ideas: In **food**, spices from one region changed diets in another. In **religion,** foreign temples and shrines drew new followers. In **clothing,** fabrics and techniques blended into hybrid styles. In **technology,** ship designs, tools, and farming methods spread across continents. Ports were schools where no authority could fully control the curriculum.

Security and Control

Wealth attracted danger. Pirates prowled sea routes. Rival empires coveted trade profits. Harbors built defenses: walls, towers, and chains that stretched across entrances. Naval fleets stationed nearby offered protection for allies and warnings to enemies.

Governments issued strict regulations to ensure accurate taxes and discourage smuggling. Inspectors tested weights, examined goods, and checked manifests.

Despite safeguards, conflict simmered. Dockworker strikes, warehouse theft, and disputes between officials and merchants marked daily life. Yet the risks were tolerated because the rewards were immense.

Knowledge Carried in the Hull

A ship entering port delivered more than goods. It brought maps, sailing instructions, stories of storms, and information about shifting winds and currents. It spread knowledge faster than any land route could.

Ports preserved that knowledge in the minds of navigators and the scrolls of scholars. Astronomers refined their understanding of stars based on reports from distant voyages. Cartographers updated coastlines and distances with each new arrival. Learning bent its shape to match maritime truth.

Social Mobility Rises with the Tide

Waterfronts offered opportunities rare inland. A skilled sailor could rise in rank. A merchant who gambled on cargo might gain fortune in one voyage. Farmers sold produce to supply hungry crews. Craftsmen built ships, barrels, ropes, and tools. Dockworkers earned wages that supported families.

Ports welcomed ambition. A foreign trader with useful knowledge could gain influence. Harbor cities often became more tolerant than inland towns because their prosperity depended on variety, not uniformity. When cities face the sea, they must also face the world.

Ports Under Pressure

Harbors could rise and fall quickly. Silt from rivers could clog entrances, forcing costly dredging or abandonment. Political turmoil could push merchants away. Earthquakes and tsunamis destroyed waterfronts in moments. After Rome's collapse, Ostia declined as trade networks shifted.

No port was guaranteed success. But while they thrived, they shaped entire civilizations.

New Reality of Movement

The docks rarely slept. Even when the sun dipped below the horizon, ships unloaded under torchlight. Night markets opened. Sailors looked for rest or company. Harbor guards patrolled the quays. The rhythm of activity beat like a pulse that kept the world alive.

These global gateways proved that distance did not require disconnection. With each arrival and departure, they stitched continents into a web of movement that changed how people understood the earth. Ports taught humanity that strangers could become partners and that foreign lands could become familiar.

Trade routes did not only bring goods—they brought the world together.

The Sea Extends a Welcome

As the merchant ship prepared to depart again, sailors stowed fresh provisions and tightened ropes. The lighthouse flame gleamed above like a guardian watching past and future. Once the tide turned, the vessel would leave the crowded harbor and disappear toward unseen markets.

Every departure carried hope of return. Every arrival brought new possibilities.

Ancient ports proved that where the sea meets land, cultures meet too—and the world grows larger with every wave.

Chapter 17
Information on the Move – Couriers, Maps, and Early Navigation Science

The mapmaker leaned over a large sheet of parchment, holding a fine brush between fingers stained with ink. Lamplight flickered against the walls of the scriptorium, casting long shadows over wooden tables strewn with compasses, wax tablets, and carved rulers. Outside, night pressed against shuttered windows, but inside, the room vibrated with quiet purpose. Scholars whispered as they compared star charts and measured angles with precision. The mapmaker paused to steady his hand, then drew a curve—another coastline taking shape, ready to guide travelers into spaces once considered unknown. An apprentice recorded distances noted by merchants and sailors, checking them against star positions marked on celestial globes. Every stroke of ink carried more than geography. It carried the hope of safe passage, accurate messages, and a clearer picture of the world.

From the earliest cities to far-reaching empires, information shaped power. To move messages reliably and understand direction with confidence meant controlling distance itself.

The Need for Reliable Knowledge

As cities expanded and rulers managed larger territories, the challenge shifted from reaching new lands to coordinating them. Roads and sea lanes helped people and goods move, but without accurate knowledge of direction, time, and distance, travel remained uncertain. Leaders needed maps grounded in measurement, not memory.

Surveyors watched how shadows changed position during the day. Scholars compared mountain peaks with the rising and setting of known stars. These methods turned vague landscapes into detailed records. Rivers stopped vanishing into guesswork. Coastlines no longer twisted randomly across parchment.

Each improved map became a quiet promise: follow this, and you will arrive. Empires depended on that promise to function.

Keeping Time with the Sun

One of the earliest navigation tools came from watching sunlight. A gnomon—a simple vertical stick—cast a shadow that moved predictably throughout the day. By observing its position, people identified cardinal directions and marked seasonal changes. Farmers planned planting and harvest. Travelers oriented themselves along roads by comparing shadow length and angle.

Over time, stone and bronze gnomons appeared in temples and city squares. Their carefully carved surfaces measured angles with surprising accuracy. On ships, portable shadow devices helped navigators hold their course after land slipped behind the horizon. The sun became a dependable guide, turning each clear day into a map.

Stars as Navigators' Companions

Night skies offered a second guide. Polaris, the North Star, served as an anchor point marking true north. By measuring its height above the horizon, sailors estimated latitude. Even before mathematics sharpened accuracy, constellations gave seafarers a sense of direction. The reappearance of certain stars signaled safe winds or the return of sailing seasons.

Astronomers charted celestial paths and preserved them in written texts. Navigation tied the sea to the heavens, linking human movement to cosmic order. Each voyage tested these observations, blending tradition, memory, and early science.

Maps Turn Memory into Knowledge

Early maps reflected stories rather than precise measurement. Unknown regions held drawings of beasts or warnings. Coastlines curved according to rumor. Over time, experience replaced imagination.

Egyptian surveyors measured fields after each Nile flood using ropes with evenly spaced knots. Greek scholars such as Eratosthenes estimated Earth's circumference with impressive accuracy by comparing shadows in distant cities at the same moment. These insights refined geographic thinking. Concepts of latitude and longitude emerged. Drawings grew less symbolic and more reliable.

Roman engineers created route maps showing distances, relay stations, and the locations of bridges and inns. Travelers no longer relied on local recollection alone. Maps shifted from decoration to dependable instruments of movement.

Paper and Ink Accelerate Communication

The spread of paper from China transformed recordkeeping. Lighter, smoother, and easier to produce than earlier materials, paper allowed rapid copying of maps, scrolls, and official orders. Courier systems immediately benefited. Messages became lighter and easier to carry. Governments could issue information more widely and more often.

Paper supported the wider distribution of knowledge. More people could read, copy, correct, and preserve information about distance, direction, and terrain.

Couriers Who Carried the World

Control over distance depended on who moved information and how quickly it reached its destination. Empires built courier networks that turned geography into manageable space.

In Persia, mounted couriers raced along the Royal Road, changing horses at waystations. Herodotus praised their speed and endurance, claiming no storm or hardship slowed them. While poetic, the system was remarkably effective.

Rome developed the cursus publicus, a state-run courier service. Riders and wagons moved along roads paved for efficient travel. Relay stations offered fresh animals and shelter. Priority signals carried the emperor's seal. Messages that once took months to travel could now arrive in weeks.

In the Andes, the Inca relied on chasqui runners who passed messages along mountain roads. They carried quipus—knotted cords recording information. Altitude and terrain did not stop them. Their lungs, legs, and training formed an extraordinary human network. Couriers became the living threads stitching distant territories together.

Math and Measurement Refine Direction

Geometry played a central role in sharpening navigation. Greek mathematicians used angles and circles to determine relative positions. By measuring a few key distances and angles, surveyors mapped entire regions through triangulation.

Time also guided navigation. Water clocks and shadow calculations allowed travelers to compare local time with known reference points, offering early methods for estimating east-west position. Each improvement replaced uncertainty with calculation.

Portable Instruments of Precision

Compact tools carried scientific knowledge into daily travel. The dioptra helped surveyors measure angles. The quadrant and astrolabe served as early analog computers. By aligning sights with stars or the sun, navigators confirmed latitude, predicted seasonal conditions, and checked their course at sea.

These instruments required training, often reserved for priests, astronomers, and professional navigators. Yet through trade and instruction, their influence spread widely.

Charts That Went to Sea

Mariners needed charts distinct from inland maps. Sea charts focused on coastlines, reefs, currents, and harbor entrances. Depth soundings marked where ships might run aground. Compass roses showed wind directions and favored headings.

These charts preserved the wisdom of countless voyages. Every number represented an experience—often a narrow escape.

Sailors grew to trust these drawings as essential companions on open water.

Writing That Shaped Rulership

Accurate information strengthened governments. Laws reached provincial officials with fewer delays. Tax records traveled securely. Reports of danger arrived before crises spread. Treaties could be upheld across long distances.

Merchants relied on written manifests and route instructions. Clear recordkeeping encouraged trust, fairness, and stable commerce. The state grew smarter because its words could travel.

Knowledge Moves with the Messenger

Maps and instruments guided movement, but human memory carried details that no parchment could capture. Couriers recalled washed-out bridges, dangerous passes, and new shortcuts. Sailors brought home fresh knowledge of currents, winds, and hazards. They shared their discoveries in taverns and shipyards, helping the next voyage succeed.

Experience and recordkeeping intertwined into a stronger system.

The Cost of the Message

Reliable information required investment. Relay stations needed maintenance. Animals demanded forage. Instruments required skilled artisans. Maps took time and expertise to produce. Governments funded these networks through taxes on trade.

Yet the benefits were enormous. Faster communication prevented rebellion, improved trade, and reduced danger. Accurate navigation saved lives, ships, and cargo. Knowledge became a form of infrastructure as vital as roads or ships.

Ideas Crossing Borders

As maps and navigation tools traveled, they carried cultural exchange. A chart copied in one port contained observations from another. A Babylonian star measurement guided a Greek navigator. Egyptian surveying methods influenced land division across the Mediterranean. Chinese innovations in paper and magnetic understanding would later reshape navigation across the world.

Armies could be stopped by borders. Information could not.

Information Remakes the World

Back in the scriptorium, the mapmaker leaned back from his parchment. The coastline stretched clearly now, no longer a rumor but a record. A region that had once been a blank space invited ships and travelers. Scholars rolled up finished scrolls and secured them in leather cases for journeys ahead. Outside, stars traced familiar patterns across the sky.

Every map meant new opportunity. Every courier carried new possibility. Tools and charts gave rulers and merchants sharper choices. Navigators trusted instruments that linked their future path to the wisdom of those who traveled before.

Distance no longer marked limitation.
With each improved direction and faster message, ancient societies discovered that movement depended not only on strength but on understanding.
Knowledge crossed deserts and oceans faster than

grain or gold. And wherever accurate information arrived, prosperity and progress soon followed.

Chapter 18
The Cost of Motion – Risks, Piracy, and the Price of Progress

Wind howled as storm clouds swallowed the horizon. Waves battered the merchant ship, splashing icy water across the deck as sailors fought to keep control. In the distance, a smaller vessel rode the swells with alarming ease. Armed men crowded its narrow hull, waiting for the storm to weaken their target. The merchant crew worked faster, hearts pounding with the knowledge that nature's fury and human greed could strike at the same time. Rain lashed their faces while the sea rose like a wall. They prayed to survive both the ocean and the danger riding its surface, knowing that travel, the source of their livelihood, carried steep risks.

Movement made the world richer and more connected. But motion always came with a price.

Storms That Ruled the Waves

Water offered speed, but it obeyed no master. Sailors studied seasons to avoid the worst storms, yet even the most experienced captain could be caught unaware. Sudden squalls in the Mediterranean shattered masts. The Indian Ocean's monsoons overturned ships that misread the winds. The Pacific tested skill with waves tall enough to hide entire fleets.

Shipbuilders strengthened hulls, but wood strained under repeated blows. Cargo shifted and overturned vessels. Sails snapped under hurricane winds. Lightning ignited tarred rigging. A single unnoticed leak could doom even a sturdy ship.

Maps marked dangerous waters where wrecks gave warning. Still, sailors returned to sea because the ocean carried the goods people needed and the fortunes traders sought. Risk became a companion as constant as the tide.

Hidden Dangers Beneath the Surface

Even on calm days, the sea hid its threats. Reefs sliced hulls like sharpened blades. Sandbars shifted with tides, trapping ships where deep water once ran. Sailors dropped lead lines to test depth, but error meant disaster. Captains often knew coastlines only through stories. One wrong turn brought ruin.

When a ship went down, rescue was rare. Survivors clung to broken planks or scrambled onto jagged rocks. Near settled shores, fishermen might intervene. On open ocean, fate decided everything. Shipwrecks entered myth and song because loss on the water became a warning for those who followed. The ocean offered no mercy to those who crossed it—and yet they crossed anyway.

Outlaws on the Waves

Where wealth traveled, predators followed. Pirates, driven by greed or desperation, prowled major sea routes. They learned which ports produced rich cargo and which capes allowed ambush. Fast, lightly loaded ships struck without warning, boarded quickly, and vanished before authorities responded.

In the Mediterranean, pirates plagued traders despite Greek and Roman efforts to stop them. Cilician pirates grew bold enough to raid Italy's coast, forcing Rome to launch a massive campaign to suppress them—at least for a while.

In the Indian Ocean, sailors feared raids near the Horn of Africa. In East Asia, the South China Sea hosted pirate fleets that rivaled small navies. They demanded ransom or killed crews to seize cargo swiftly. Fear was their most effective weapon.

Pirates thrived in the gaps between states, along coastlines without strong rulers, and in waters where storms scattered patrols.

Guarding Against Attacks

To counter piracy, states built warships and patrolled shipping lanes. Naval escorts sometimes joined merchant convoys, each ship watching the horizon. Archers stood ready behind railings. Soldiers defended elevated decks to prevent boarding.

Coastal fortresses monitored sea routes, signaling danger with fire or flags. Harbors stretched heavy chains across entrances to block hostile ships. Hunters pursued the hunters, but no system proved perfect. As long as ships carried valuable cargo, someone would try to take it.

Bandits on the Land

Travel by land carried its own dangers. Bandits hid in forests or rocky passes, waiting for caravans to falter. A broken wheel or a lame horse created opportunity. Thieves struck quickly, knowing guards could not protect every wagon.

Mountain routes posed special threats. Narrow roads funneled travelers into predictable paths. Lookouts signaled from high cliffs. Ambushes unfolded with ruthless coordination. By the time help arrived—if it did—bandits vanished into the landscape.

Even peaceful roads became hazardous when law enforcement weakened. Merchants hired guards or traveled in large groups. Safety in numbers became a rule every traveler followed.

Disease: The Unseen Enemy

Mobility spread ideas and goods, but it also spread illness. Traders returning from distant ports carried more than spices or silk—they sometimes carried contagion no one had encountered before.

In crowded caravansaries and waterfront inns, travelers gathered in tight quarters. Fleas clung to animals. Contaminated water spread sickness. A single cough could infect an entire ship.

Some ports enforced quarantine, isolating vessels before allowing people ashore. But many communities lacked knowledge or discipline. The same motion that enriched societies threatened them with biological disaster. The toll of progress was measured not only in money, but in lives.

Taxes and Tolls: Protection or Profit?

Governments claimed responsibility for securing roads and sea routes. To fund patrols and defenses, officials charged tolls on bridges, gates, and ports. Caravans paid customs dues at borders. Some fees supported real improvements—lighthouses, guards, and repairs.

But taxes also became burdens. Corrupt officials invented new charges or demanded bribes. Travelers resented paying for protection they never saw. Merchants grumbled that as soon as trade flourished, rulers reached deeper into their purses. Mobility gave rulers opportunity. Some used it wisely. Others used it greedily.

Insurance Before Its Name

Merchants sometimes divided goods among multiple ships so that one loss did not ruin everything. Partnerships formed in markets where traders agreed to share losses, knowing that mutual aid increased everyone's chances of survival.

These practices foreshadowed early insurance—communities learning to manage danger collectively rather than relying on luck.

Technology Answers the Threat

Every danger pushed innovation. Stronger hulls reduced wrecks. Improved road design made ambushes harder. Navigation tools helped sailors avoid reefs. Warships evolved to protect trade and combat piracy. Metal weapons discouraged bandits. Even writing played a role—documents identified legitimate travelers and reduced fraud. Fear often accelerated creativity.

Diplomacy Over the Waves

Because trade connected nations, threats to trade sometimes forced cooperation. States negotiated agreements promising mutual defense of sea routes. Fleets from one kingdom patrolled coasts near another. Sea treaties established safe passage zones and punished violators.

Piracy could unite rivals. Shared danger reminded rulers that wealth flowed more smoothly when nations worked together, even briefly. Diplomatic solutions appeared because all sides benefited when goods arrived safely.

Conflict Born from Opportunity

Movement reshaped power. Ports grew wealthy, attracting rivals. Roads helped armies strike faster. Control of a strait meant control of entire markets. These advantages tempted conflict.

Carthage and Rome fought for Mediterranean dominance. Greek city-states battled for sea lanes that fed their people. Chinese dynasties and nomadic tribes clashed over Silk Road influence. Where movement brought prosperity, it also brought rivalry.

Responsibility and Risk in Equal Measure

Travelers accepted danger because reward made the risk worthwhile. A successful journey meant profit, knowledge, and prestige. Losing everything meant sorrow and ruin. Merchants weighed these outcomes every time they set out.

Governments faced similar choices. Neglecting security could collapse economies. Overregulating movement could suffocate trade. Survival depended on balance.

More Movement, More Vulnerability

The very networks that strengthened societies also made them more fragile. When a bridge collapsed or a port closed, markets faltered. When piracy increased, trade slowed. When roads deteriorated, messages arrived too late. Interconnectedness helped civilizations grow—and made them easier to disrupt.

When people depend on motion, stopping becomes the most powerful threat of all.

Courage on the Journey

Back on the storm-lashed deck, sailors tightened ropes with trembling hands. Lightning carved white scars across the sky. The smaller pirate vessel fought the same waves, its crew determined to reach a vulnerable target. Fear narrowed the world to wind, water, and the glint of weapons in the dark.

Storm or bandit, disease or toll collector—danger followed those who dared to move. Yet every day, ships set sail and caravans marched. Travel shaped empires and lives, proving that the rewards outweighed the risks for those bold enough to continue.

The cost of motion never disappeared. It fueled invention, encouraged cooperation, and revealed the fragile line between prosperity and peril. People traveled anyway, because staying still meant surrendering possibility. Movement brought risk. It also brought the future.

Chapter 19
When Roads End – The Decline of Ancient Mobility Networks

Grass pushed through cracked stone as the road disappeared into overgrown brush. Fallen milestones lay half-buried in soil. Where caravans once carried spices, metals, and news of distant lands, silence now ruled. A few scattered travelers moved cautiously along this forgotten path, stepping around broken paving and washed-out culverts. Villages nearby no longer welcomed visitors from afar. Children grew up knowing only the horizon they could walk to before nightfall. The road that once united worlds had faded into memory.

Networks of movement made ancient civilization thrive. Their collapse reshaped the world just as powerfully.

When Empires Fall, So Do Their Roads

Roads and ports thrived when central governments collected enough taxes to maintain them. Armies kept peace. Bureaucrats coordinated repairs. Merchants invested in travel because laws protected their goods. But when empires weakened, the roads beneath their feet suffered first.

Political fragmentation disrupted order. Wars consumed resources once used for upkeep. Local rulers prioritized defense over mobility. Tax collectors vanished or demanded payments without providing services. Bridges rotted. Harbors silted over. Relay stations closed. No one dredged shipping channels or organized wagon repairs. Without regular attention, infrastructure crumbled quickly.

Movement, once seamless, became an exhausting risk.

Rome: From Highways of Power to Paths of Survival

The Roman Empire reached its height with a vast system of roads stretching across Europe, North Africa, and the Near East. These routes allowed soldiers to march, merchants to sell goods, and messages to travel with remarkable speed.

Then pressures mounted. Invading armies cut routes. Internal conflict drained the treasury. Local elites drifted from imperial authority. With no stable funding, repair crews disappeared. Bridges collapsed under winter ice or floods and were never rebuilt. Trade faltered. Cities shrank. The saying "All roads lead to Rome" remained true in memory but faded in practice.

By the early Middle Ages, many roads served only local traffic. Farmers guided carts over grass-covered foundations laid centuries earlier. Wealth narrowed, and people lived more of their lives without ever seeing a market beyond their nearest town.

Persia: Crossroads Disrupted

The Persian Royal Road connected distant regions with impressive efficiency during Achaemenid times. Relay stations buzzed with activity as couriers linked the empire's farthest corners.

After Alexander's conquests and later political shifts, Persia fractured into smaller states. Some rulers maintained portions of the old network, but deserts crept over abandoned stretches. Conflict redirected commerce to different paths. Caravans faced greater danger from raids as security declined.

Markets that once thrived with international exchange turned inward. Knowledge that had flowed through Persia's heart slowed to a trickle.

China: The Rhythm of Rise and Decline

China experienced cycles of unity and division. When strong dynasties ruled, roads, canals, and postal lines flourished. The Silk Road linked China to Central Asia and beyond. But when dynasties fell, fragmentation followed. Banditry surged. Local lords blocked roads with checkpoints or fortifications.

The collapse of the Han Dynasty brought deep disruption. Educated travelers no longer reached distant courts. Taxes could not be collected reliably, so infrastructure withered. Some routes vanished into wilderness until later dynasties restored them at great cost. Decline slowed progress, but recovery remained possible when unity returned.

Ports Abandoned to the Tide

Harbors changed as well. When governments could no longer maintain lighthouses or dredge channels, ships found safer destinations. Trade shifted elsewhere. Ports that once thrived decayed under salt wind.

Warehouses emptied. Customs officials left or turned to corruption. Stranded communities lost connections to distant markets. Societies once rich in imported goods and ideas became confined to what could be grown or made locally. Innovation, once driven by global exchange, became a lonely effort.

Couriers Without Cause

Communication systems designed to move messages fast disappeared with the institutions that supported them. Without standardized routes, messengers lost stations where they could rest or change mounts. Travel became unpredictable.

Information slowed sharply. Decisions lagged. Leaders ruled without knowing what happened beyond their sight. Regions drifted apart as misunderstanding grew. The world became larger again, not in distance but in separation.

Knowledge Trapped in Place

With less travel came fewer encounters with new ideas. The disappearance of traders meant tools and inventions spread slowly, if at all. Books were lost in fires and wars. Libraries fell into ruin. Scholars feared journeys that once inspired them. The decline of movement turned curiosity into risk.

Infrastructure as Memory

Collapsed roads and bridges lingered as reminders of achievements now out of reach. Moss covered carved stones. Weeds reclaimed embankments. Families walking these remnants heard stories of a time when distant lands felt close.

Technology survived in isolated pockets. A lone shipwright remembered how to build a strong hull. A map survived in a monastery. Some knowledge endured because individuals refused to let the world shrink completely. Humanity did not forget entirely—but it remembered with longing.

Economies Turn Inward

When movement stopped, specialization shrank. A village that once imported metal tools learned to live without them. Farmers grew only what they could consume or haul short distances. Markets narrowed. Wealth distributed unevenly.

Those far from remaining strongholds of commerce suffered most. The edge of the empire became the center of hardship.

Conflict Replaces Commerce

Without stable networks, violence replaced agreement. Groups fought for supplies now difficult to obtain. Rival settlements built fences, ditches, and barricades. Roads that once united became contested lines rather than shared assets.

Warriors stood where merchants once bargained. People protected what little they had instead of seeking more through exchange. Global links collapsed into tangled threads.

Isolation as a New Reality

Generations grew up never seeing silk or spices. Coins with foreign inscriptions became curiosities instead of currency. Knowledge of faraway regions faded. Maps returned to blank spaces and imagined dangers.

Community identities hardened. Stories of outsiders shaped fear. Language drifted as neighbors met too rarely to share words. The world fractured into smaller worlds. Distance regained its power to divide.

What Decline Teaches

The fall of ancient mobility networks reveals an essential truth: infrastructure is not permanent. It demands attention and collective effort. When societies turn inward or become divided, roads and ports suffer first—and everything else follows.

Repairing what is lost can take centuries. The lessons of movement fade when unused. Skills vanish faster than stone decays. Movement is civilization's heartbeat. When it slows, the entire body weakens.

The Road Waits for Return

Late afternoon sunlight cut through broken branches to reveal a fragment of paving beneath the grass. A child might trace the pattern of stones and wonder what lies beyond the nearest hill. Curiosity remains even when roads fail.

Civilizations rise and fall. But the urge to connect, to explore, to trade and communicate does not die. In later ages, new rulers would clear old pathways or carve new ones, rediscovering that strength lies in movement.

When roads end, decline begins. But every forgotten road holds a quiet promise: travel will someday begin again.

Chapter 20 Epilogue: The Eternal Journey – How Connection Continues to Shape Civilization

From thousands of miles above Earth, a satellite camera captures the glow of cities and the faint threads that bind them together. Highways twist through mountains and deserts. Container ships cross dark oceans guided by precise digital signals. Aircraft trace arcs through thin skies. Yet look closer and another story appears beneath the modern map: ancient roads still influencing today's paths, old harbors guiding new ports, and forgotten caravan trails forming the hidden framework for global movement. Across the continents, the work of long-gone engineers remains visible to anyone who knows where to look. Ancient grooves in stone lie parallel to the smooth asphalt of modern highways. Breakwaters built centuries ago dictate how container ships line up to dock. The world still follows the routes carved by our ancestors.

Transportation today seems fast enough to outrun history, but the truth is that we move along channels first shaped in distant times. The future does not erase the past. It rides upon it.

Movement Defines Civilization

From the earliest cities to the sprawling nations of today, societies thrive when people can travel, trade, and communicate. The idea is simple: connection creates opportunity. Long before engines roared or steel rails crossed continents, ancient builders understood this truth. Their roads and ships, their maps and compasses, created a system that let ideas and goods flow farther than any single village could imagine.

Modern globalization has roots in those early networks. A Roman road beneath a European street still influences where buildings stand. A port built in the Mediterranean two thousand years ago remains active because the bay offers shelter no modern engineer can improve. Technology changes, but geography and ancient ingenuity endure.

Ancient Foundations, Modern Expansion

Innovation rarely begins from nothing. Today's infrastructure reflects ancient ingenuity in countless ways.

Roads still rely on principles developed by the Romans. Engineers build layered foundations, manage drainage, and curve surfaces to carry water away. The bones of Roman engineering helped shape motorways that stretch across continents.

Maritime trade rests on knowledge inherited from early explorers. Wind patterns, sea currents, and seasonal shifts continue to guide shipping schedules. The monsoon routes that once carried dhows between Arabia and India now direct steel ships hauling millions of tons of cargo.

Navigation systems depend on mathematics perfected by scholars who mapped coastlines under lamplight. Satellites replaced stars, but the goal remains the same: know where you are and where you are headed.

Global shipping lanes follow pathways first traced by merchants traveling with wind and courage. Ports built where natural harbors offered refuge have grown into terminals where cranes tower higher than lighthouses once did. The ancient world planted the seeds that grew into modern networks.

Knowledge Inherited Across Time

Much of what keeps civilization moving today comes from discoveries long ago. Magnetic compasses, invented and refined in China, remain essential to maritime navigation. The concept of mapping a sphere emerged from Greek mathematical insight and shaped global geography.

Standardized measurement systems and relay communication networks that once served empires evolved into postal services and digital routing. Where horse hooves once echoed on stone roads, fiber-optic cables now pulse with light. Both serve the same purpose: keep people connected.

Ancient infrastructure taught the world that distance can be conquered through planning, cooperation, and belief in improvement.

The Heartbeat of the Global Economy

Modern trade moves at astonishing speed. Cargo ships carry clothing from Southeast Asia to markets in Europe. Food from South America fills tables in North America. Oil from the Middle East powers vehicles in Africa and Asia. Airplanes deliver fresh goods across oceans in hours. Data crosses the globe instantly, allowing markets to respond faster than storms can form.

This global network has created prosperity on a scale ancient builders could not have imagined. But they would recognize the principle: movement creates wealth. Goods must reach buyers. People must reach opportunities. Ideas must reach open minds.

The same incentives that built the Royal Road, Roman highways, and Silk Road corridors still drive construction today. Where there is commerce, there will be networks. Where there are networks, civilization will grow.

Challenges That Echo Across Centuries

Despite modern achievements, many challenges mirror those faced long ago.

Storms still threaten ships. Earthquakes and erosion challenge bridges and tunnels. Piracy, though less common, remains in some regions. Cyberattacks target digital trade routes much like bandits once targeted desert caravans. Governments continue to tax transportation, sometimes to fund protection, sometimes for revenue alone.

And when systems fail—through war, disaster, or neglect—the consequences are familiar: isolation, economic strain, and uncertainty.

History reminds us that infrastructure requires care. The fall of ancient mobility networks warns that progress is never permanent.

Technology Shrinks Distance Further

The ancient world reduced barriers so people could walk or sail to new places. Modern tools push that idea outward.

Automobiles turned daily life into a journey. Railways sped travelers across land at rates no horse could match. Jet engines collapsed oceans into overnight flights.

Now digital networks allow information to move faster than any physical traveler. Video calls erase borders. Satellites monitor storms and guide vehicles with remarkable precision. Space exploration hints at future routes extending beyond Earth.

Each breakthrough continues the same story: distance becomes smaller when humans refuse to accept limits.

The Human Desire to Connect

What drove ancient builders remains alive today. Curiosity inspires exploration. Trade encourages cooperation. Families stretch across borders and maintain ties. Students travel to learn. Pilgrims journey to sacred places. Tourists seek wonder in faraway landscapes.

Mobility reveals the shared threads of humanity. In crowded stations or airports, the mix of travelers shows how connection brings worlds together. To move is to live among others.

The Past Beneath Our Feet

Imagine standing on a modern highway bridge. Beneath the concrete lie layers of history. A Roman engineer might recognize the same alignment chosen long ago. Centuries later, a medieval caravan could have crossed nearby with donkeys and camels. Now automobiles speed over the same ground.

Technology evolves. Purpose persists. The sea tells the same story. Today's supertankers dock in bays once visited by wooden ships with hand-stitched sails. Fishermen from thousands of years ago would still recognize the rhythm of tides and winds that govern entry to safe harbors.

Humanity travels forward while looking back, whether consciously or not.

Legacy and Responsibility

Understanding how ancient infrastructure shaped the modern world creates a responsibility to build wisely. Roads must withstand climate change and stronger storms. Ports must expand to handle increasing trade while protecting fragile coastlines. Communication networks must resist disruption and remain accessible to communities that rely on them.

We must remember why networks exist: to serve people. When roads crumble or ports close, communities lose connection and hope. Ancient decline taught that neglect can unravel centuries of progress.

The future depends on maintaining the flow of knowledge, goods, and relationships. No technology protects against complacency.

Continuing the Eternal Journey

The satellite image updates again, showing planes rising from runways and ships leaving harbor. A train crosses a river on a bridge first imagined by minds who feared the water's depth but built anyway. Trucks follow paths once walked by Romans and Persians. Data signals leap between towers, echoing the message relays of ancient horsemen.

Movement continues.

People build because they want to reach one another. They plan journeys because discovery brings growth. They take risks because opportunity lies beyond the horizon. The ancient world proved that civilization advances through connection—and declines when it stops.

Humanity's greatest invention is not the wheel or the sail or the compass. It is the network they created together.

The Story of Transportation Is the Story of Us

We are a traveling species. Our history is written on roads, traced through sea routes, and measured in the distances we have conquered. Every step farther has made the world richer in ideas and relationships.

Ancient innovations formed the first chapters. Today continues the narrative. Tomorrow will add new routes—perhaps in the skies above or on the planets beyond. But wherever the journey leads, the goal remains unchanged: to reach one another. Civilization moves forward because people refuse to stand still.

Made in the USA
Monee, IL
04 November 2026

67384803R00098